Oil Spill Occurrence, Simulation, and Behavior

Fuels and Petrochemicals

Series Editor
M.R. Riazi

Biofuels Production and Processing Technology
M.R. Riazi and David Chiaramonti

Coal Production and Processing Technology
M.R. Riazi and Rajender Gupta

Oil Spill Occurrence, Simulation, and Behavior
M.R. Riazi

For more information about this series, please visit: https://www.routledge.com/
Fuels-and-Petrochemicals/book-series/CRCFPS

Oil Spill Occurrence, Simulation, and Behavior

M.R. Riazi

CRC Press
Taylor & Francis Group
Boca Raton London New York

CRC Press is an imprint of the
Taylor & Francis Group, an **informa** business

First edition published 2021
by CRC Press
6000 Broken Sound Parkway NW, Suite 300, Boca Raton, FL 33487-2742

and by CRC Press
2 Park Square, Milton Park, Abingdon, Oxon, OX14 4RN

© 2021 Taylor & Francis Group, LLC
CRC Press is an imprint of Taylor & Francis Group, LLC

The right of M.R. Riazi to be identified as author of this work has been asserted by him in accordance with sections 77 and 78 of the Copyright, Designs and Patents Act 1988.

Reasonable efforts have been made to publish reliable data and information, but the author and publisher cannot assume responsibility for the validity of all materials or the consequences of their use. The authors and publishers have attempted to trace the copyright holders of all material reproduced in this publication and apologize to copyright holders if permission to publish in this form has not been obtained. If any copyright material has not been acknowledged please write and let us know so we may rectify in any future reprint.

Except as permitted under U.S. Copyright Law, no part of this book may be reprinted, reproduced, transmitted, or utilized in any form by any electronic, mechanical, or other means, now known or hereafter invented, including photocopying, microfilming, and recording, or in any information storage or retrieval system, without written permission from the publishers.

For permission to photocopy or use material electronically from this work, access www.copyright.com or contact the Copyright Clearance Center, Inc. (CCC), 222 Rosewood Drive, Danvers, MA 01923, 978-750-8400. For works that are not available on CCC please contact mpkbookspermissions@tandf.co.uk

Trademark notice: Product or corporate names may be trademarks or registered trademarks and are used only for identification and explanation without intent to infringe.

Library of Congress Cataloging-in-Publication Data

Names: Riazi, M. R., author.
Title: Oil spill occurrence, simulation, and behavior / by M.R. Riazi.
Description: First edition. | Boca Raton, FL : CRC Press/Taylor & Francis
Group, LLC, 2021. | Series: Fuels and petrochemicals | Includes
bibliographical references and index. | Summary: "Offering a unique and
comprehensive view on the latest technologies related to oil spill
occurrence, behavior, and modeling, this work provides practical insight
for professionals and academics. It discusses various sources of oil
spills, major accidents including the 2010 Gulf of Mexico oil spill, and
modeling and simulation techniques to predict rate of oil spill
disappearance. It also covers characteristics of crude oil and its
products as well as clean-up methods"-- Provided by publisher.
Identifiers: LCCN 2020048955 (print) | LCCN 2020048956 (ebook) | ISBN
9781138362307 (hardback) | ISBN 9780429432156 (ebook)
Subjects: LCSH: Oil spills--Simulation methods.
Classification: LCC TD196.P4 R53 2021 (print) | LCC TD196.P4 (ebook) |
DDC 363.738/2--dc23
LC record available at https://lccn.loc.gov/2020048955
LC ebook record available at https://lccn.loc.gov/2020048956

ISBN: 978-1-138-36230-7 (hbk)
ISBN: 978-0-429-43215-6 (ebk)

Typeset in Times
by Deanta Global Publishing Services, Chennai, India

To:

My family, the memory of my parents, my former professors, colleagues, and the world scientific community

Contents

Preface xi
Author Biography xiii

Chapter 1 Introduction 1

References 14

Chapter 2 Characteristics and Properties of Crude Oil and Its Products 17

2.1 Nature of Crude Oils and Their Products 19
2.2 Basic Properties of Pure Hydrocarbons 23
2.3 Basic Properties of Crude Oil and Its Products 26
2.4 Estimation of the Composition of Petroleum Mixtures 32
2.5 Estimation of Density and Pour Point 33
2.6 Estimation of Boiling Point, Flash Point, and Vapor Pressure 37
2.7 Estimation of Viscosity, Diffusivity, and Surface Tension 38
2.8 Solubility of Hydrocarbons and Petroleum Fluids in Water ... 42
2.9 Summary 45
References 46

Chapter 3 Sources and Causes of Oil Spills 47

3.1 Introduction 47
3.2 Statistical Data on Oil and Gas Production and Trade
Movement 49
3.3 Oil Spills Caused by Petroleum Exploration and Production 55
3.4 Oil Spills Caused by Marine Transportation 59
3.4.1 Oil and Gas Transportation by Tankers 60
3.4.2 Oil and Gas Transportation by Pipelines 65
3.5 Oil Spills Caused by Refineries 69
3.6 Oil Spills Caused by Natural Seeps 69
3.7 Oil Spills Caused by War, Sabotage, and Other Actions 71
3.8 Summary 76
References 77

Chapter 4 Major Oil Spills: Occurrence and Their Impacts 81

4.1 Introduction 81
4.2 Oil Spill Detection 84
4.3 Major Accidents 85
4.4 Tanker Oil Spills 90
4.4.1 *Sanchi* Oil Spill 90
4.4.2 *Haven* Oil Spill 92

vii

viii   Contents

| | 4.4.3 | *Exxon Valdez* Oil Spill | 93 |

4.5 Persian Gulf Oil Spill 95
4.6 Deepwater Horizon (Gulf of Mexico) Oil Spill 2010 98
    4.6.1 Chronology of Events 99
    4.6.2 How it Occurred 102
    4.6.3 Deepwater Horizon, Macondo Well, and Fluid
          Composition 115
    4.6.4 Meteorological Data 118
    4.6.5 Oil Flow Rate and Slick Area 122
          4.6.5.1 Oil Flow Rate 122
          4.6.5.2 Surface Area 122
    4.6.6 Trajectory and Simulation 126
4.7 Summary 131
References 131

Chapter 5 Environmental, Economic, and Political Impacts: Case of BP
Oil Spill 135

5.1 Introduction 135
5.2 Environmental Impacts 136
    5.2.1 Impacts on the Birds and Fish 136
    5.2.2 Impacts on the Shores and People 140
5.3 Economic Impacts 149
5.4 How the Oil Spill Was Stopped 151
5.5 Brief History of BP 155
5.6 Political Impacts 157
5.7 Summary 162

Chapter 6 Oil Spill Response and Cleanup Methods: Case of BP Oil Spill 163

6.1 Introduction 163
6.2 Natural Processes 165
    6.2.1 Evaporation 166
    6.2.2 Spreading 167
    6.2.3 Dispersion 167
    6.2.4 Dissolution 167
    6.2.5 Emulsification 169
    6.2.6 Oxidation 169
    6.2.7 Sedimentation 169
    6.2.8 Biodegradation 170
6.3 Protective Booms 171
6.4 In-Situ Burning 175
6.5 Dispersants 181
6.6 Skimming and Mechanical Removal Methods 189
6.7 Summary 195
References 197

Contents   ix

Chapter 7   Simple Models to Predict Rate of Oil Spill Disappearance 199

  7.1   Introduction 200
  7.2   Modeling Scheme 204
  7.3   Calculation Procedure 207
  7.4   Laboratory Experiments and Model Predictions 211
  7.5   Application of the Model to the Case of Continuous Flow
        of Oil 217
  7.6   Summary 220
  References 222

Chapter 8   Advanced Oil Spill Modeling and Simulation Techniques 225

  8.1   Introduction 225
  8.2   Near-Field and Far-Field Oil Spill Modeling 226
  8.3   Subsea Blowouts and Near-Field Modeling 227
        8.3.1   Introduction 227
        8.3.2   Reservoir Fluid versus Surface Oil 228
        8.3.3   Modeling the Underwater Oil Jet 228
                8.3.3.1   The Physical Processes 228
                8.3.3.2   Current Modeling Approaches 230
                8.3.3.3   Equilibrium Droplet Size Model 231
                8.3.3.4   Population Dynamic Droplet
                          Size Model 233
                8.3.3.5   Modeling Oil Plume Dynamics and
                          Underwater Degradation Processes 234
                8.3.3.6   Typical Near-Field Model Inputs and
                          Produced Results 241
  8.4   Far-Field Oil Spill Modeling 242
        8.4.1   Oil Spill Advection 244
        8.4.2   The Importance of Wave Data 249
        8.4.3   Other Critical Inputs and Processes 249
                8.4.3.1   Oil Entrainment and Droplet Size
                          Distribution 250
                8.4.3.2   Oil Spill Weathering and Oil Decaying
                          Processes 251
        8.4.4   Modeling of the Response Measures 256
        8.4.5   Statistical Analysis for Contingency Planning 256
  8.5   Summary 260
  8.6   Acknowledgments 261
  References 262

Chapter 9   Economic and Financial Impacts of Oil Spills 265

  9.1   Oil, Oil Spill, Response and Recovery System 265
        9.1.1   Oil and Gas System and Potential Oil Spills 266
                9.1.1.1   Oil and Gas System 266

x | Contents

		9.1.1.2	Potential Oil Spills	267
	9.1.2	Oil Spill Response and Recovery System		268
9.2	Economic and Financial Aspects of Oil Spill			269
	9.2.1	Local Economic Impact		271
		9.2.1.1	Cleanup Costs	271
		9.2.1.2	Natural Resource Damages	280
		9.2.1.3	Socioeconomic Loss	283
		9.2.1.4	Other Costs	284
	9.2.2	Global Economic Impact		285
9.3	Major Oil Spills: Compensation, Claim, and Cost Estimation			285
9.4	Summary			290
References				290

Index 293

Preface

The main causes of oil spill occurrence are accidents happening during offshore oil production activities and the transportation of crude oil, its liquid products, and liquefied natural gas (LNG) by sea through tankers. Currently, offshore production accounts for about 30% of total world oil production, while the annual amount of crude oil, products, and LNG transported by sea is about 3 billion tons and growing. A giant supertanker is capable of carrying about 2 million barrels of crude oil. In the US alone about 137 oil spill occurrences were recorded by NOAA for the year 2018. One of the major oil spill accidents occurred as a result of an explosion on the Deepwater Horizon oil rig in the Gulf of Mexico (GOM) during offshore production in April 2010. It was the largest oil spill in history after the Persian Gulf oil spill following the Gulf War in 1991.

The idea behind writing this book at this time was to mark the tenth anniversary of the GOM oil spill in April 2020 which was discussed with Allison Shatkin, the Managing Editor of CRC/Taylor & Francis publishing group, during the Annual AIChE Meeting in November 2017 in Minneapolis. This is a relatively short book which discusses the occurrence of oil spills, major accidents and the causes, the environmental, economic, and political impacts of the BP oil spill, response and cleanup methods, and simple and advanced modeling of oil spill behavior as well as an introduction to some oil spill simulators. Much of Chapters 4, 5, and 6 are devoted to the GOM oil spill as the greatest environmental tragedy in US history, and one chapter is devoted to the characteristics and properties of crude oil and various products which are needed to predict the behavior of an oil spill. Chapters 3 and 4 were prepared during the global lockdown due to the coronavirus pandemic which greatly impacted world oil production and consumption. However, all data related to production/consumption and accidents are from before the 2020 outbreak. The book is intended for practicing engineers, environmentalists, and scientists dealing with such environmental issues.

In completing this book, I am particularly indebted to the invited authors for their contribution: Konstantinos Kotzakoulakis and Simon C. George for their contribution of Chapter 8 on advanced simulation and modeling, and Liangliang Lu and Kujala Pentti for Chapter 9 on the cost of environmental impacts and cleanup operations. In addition, the work of Konstantinos Kotzakoulakis and Touraj Riazi, who enthusiastically agreed to take the task of reviewing some chapters with constructive comments and suggestions, is greatly appreciated. Many scientists have contributed to the science of oil spill simulation and behavior; their work has been used and acknowledged. Some brave journalists risked their lives to gather information about the oil flowing deep under the water surface; their contributions as well as those from the news networks covering the 2010 BP oil spill in the Gulf of Mexico are greatly acknowledged throughout the book, and we apologize for those references that may have been missed or not cited properly. The book is dedicated to all these people whose work and efforts have resulted in a better understanding of oil spill and its behavior, simulation, and cleanup methods.

I am also thankful to Taylor & Francis for the publication of this book and in particular to Allison Shatkin, the Managing Editor, who was helpful in every stage of this project, and her initial contacts and encouragement were instrumental to beginning this book project. The work of Gabrielle Vernachio, the Editorial Assistant and Project Coordinator at CRC, in reviewing and editing this manuscript is also appreciated. Many thanks to Keith Arnold for his work on typesetting of the manuscript and his commitment to this book project. Finally and most importantly I am thankful to the members of my family Shiva, Touraj, and Nazly whose love and support made it possible for me to complete this book.

M. R. Riazi
Montreal, Canada
July 15, 2020

Author Biography

Prof. M. R. Riazi has a doctorate and MSc degrees in chemical engineering from Pennsylvania State University and a BSc degree in chemical engineering with the highest honor from Arya-Mehr (now Sharif) University of Technology. He is the author/co-author of about 150 publications, including 8 books in the areas of energy and environment, oil, gas, coal, and biofuel properties, characterization, production, and processing, alongside 10 book chapters, including 3 chapters of the *API Technical Data Book – Petroleum Refining*. He has presented at over 100 conferences and conducted around 100 invited lectures and workshops for the petroleum and energy industry in more than 40 countries. He has served as an Assistant Professor at Penn State as well as a Visiting Professor in the Departments of Chemical and Petroleum Engineering at the following universities: Illinois/Chicago, Texas/Austin, Norwegian University of Science and Technology (Trondheim), McGill (Montrèal), University of Waterloo, Wright State (Dayton), Sharif University, IIT (India), and Kuwait University. He is also the owner and director of a petro-gas consulting firm based in Montrèal, Canada and was previously the elected Chairman of the Chemical Engineering Department at Kuwait University. During the past few decades, he has worked on projects funded by organizations such as the American Petroleum Institute (API), US Department of Energy, US National Science Foundation, GRI/Chicago, NSERC (Canada), Petrofina (Brussels), and SINTEF (Norway), as well as KPC, KFAS, KISR, KU, and KOTC (Kuwait). Prof. Riazi is currently a consultant to Kuwait Oil Company (KOC) for a major flow assurance project and was also an invited consultant at a number of European, American, and Middle Eastern oil- and energy-related companies for more than three decades. He has chaired many international conferences in petroleum and chemical engineering and is the founder and chair of the ICOGCT conference series. He has been a session chair (or co-chair) and a moderator in many international conferences, including F&PD Group Chair of AIChE and a moderator at the World Economic Forum. He is the founding editor of a book series on fuels and petrochemicals by CRC/Taylor & Francis publishing group. He was an editor for ASTM books on petroleum properties, production, and processing and conducted short courses on these topics for ASTM International. Prof. Riazi is the founder and Editor-in-Chief of *IJOGCT* as well as editorial associate of *JPSE* for the past 15 years. He is currently a Life Fellow of AIChE and an elected director of its Fuel and Petrochemical Division and a past member of SPE, ACS, and CIC.

As a result of his work, Prof. Riazi has been honored as the recipient of the following awards: Diploma of Honor from the U.S. Petroleum Engineering Society for the Outstanding Contributions to the Petroleum Industry (1996), KU Outstanding Research and Teaching Awards, and an elected Fellow of AIChE in 2013. He is a registered and licensed professional engineer in Ontario, Canada.

1 Introduction

An oil spill generally is defined as an accidental release of oil into seawater from a tanker, offshore production activity, or underground pipeline. The French equivalent for oil spill is *marée noire*, and the Spanish equivalent is *derrame de petróleo*. The oil is generally crude oil and its liquid petroleum products (such as LPG, gasoline, jet fuel, etc.) and liquefied natural gas (LNG) as well as liquid chemicals and biofuels. However, in this book, oil spill refers to crude oil and its product fuels such as liquid fuels from petroleum when floating on the seawater surface. Offshore production nearly accounts for 30% of global oil production (about 27 million barrels in 2016), and it is growing. For example, in Brazil, offshore production grew by 58% from 2005 to 2015 as reported by the IEA (2018).

The export and import of oil and its products by sea account for nearly 30% of global seaborne trade. Large crude carriers (LCC) may carry more than 2 million barrels (84 million gallons) of crude oil. The size of tankers is usually measured in terms of dead weight tons (dwt), and one ton of crude oil with API gravity of 30 is equivalent to about 7.2 barrels of oil. Thus a tanker with size of 300,000 dwt may carry about 2 million barrels of such crude oil. When the amount of oil released is greater than 700 tons, the oil spill is considered a large spill. An oil tanker passing through Strait of Hormuz in December 2018 is shown in Figure 1.1 (Russia Today, 2019).

The amount of oil transported by sea was multiplied by 27 between 1935 and 2012 (Planete Energies, 2015). An oil tanker was set on fire by a torpedo in the Sea of Oman due to political conflict in the Middle East as shown in Figure 1.2 (BBC News, 2019). Two oil tankers near the strategic Strait of Hormuz were reportedly attacked on Thursday, June 13, 2019, an assault that left one ablaze and adrift as sailors were evacuated from both vessels and the US Navy rushed to assist amid heightened tensions between Washington and Tehran (AP).

In 1991 about one million barrels of oil were released from the ship Haven into the sea about 7 miles off the coast of Genoa in Italy when an explosion occurred due to human error during the loading of the oil tanker as shown in Figure 1.3. The effect of this accident on the ecosystem was huge and could be seen even after 25 years (Haven, 1991).

The biggest oil spill in history occurred in 1991 during the Kuwait–Iraq war that poured between 5 and 8 million barrels of oil into the waters of the Persian Gulf. A second major oil spill occurred in April 2010 at the Gulf of Mexico, also known as the BP or Deepwater Horizon oil spill (Figure 1.4). In this accident between 3 and 5 million barrels of oil flowed into water from the damaged Deepwater Horizon for a period of 87 days covering an area of more than 100,000 km^2 with a total cost of about 60 billion dollars (Deepwater Horizon Report, 2011).

The main purpose of this book is to provide an update on the sources and causes of major oil spill occurrence and their environmental impacts, simulation and

FIGURE 1.1 An UK oil tanker passing through the Strait of Hosmuz (July 25, 2019) Middle East Online (MEO).

FIGURE 1.2 An oil tanker attacked.

modeling, cleanup methods, oil specifications and properties, economy and costs as well as their ecological impacts. The book has nine chapters including this introductory chapter covering these topics. Due to the importance of the specifications and properties of crude oil and its products, Chapter 2, following this introductory chapter, is devoted to topics related to the physical and thermodynamic properties of oils needed for predicting the behavior of an oil spill floating on seawater.

Introduction 3

FIGURE 1.3 Milan, Italy, 1991 oil spill (Haven, Science, 1991).

In Chapter 2 the nature and characteristics of different oils are discussed with the properties affecting an oil spill and its simulation and modeling. To simulate the fate of an oil spill and the rate of its disappearance, at least the following properties and specifications are required (Villoria et al., 1991, Riazi and Al-Enezi, 1999, Riazi and Edalat, 1996):

- Characterization of crude oils and petroleum fractions
- Pour and flash points
- Density and solubility parameters and vapor pressure
- Transport properties such as viscosity, diffusion coefficient, and surface tension

The characterization of crude oil and petroleum products plays an important role in using laboratory data to define such complex mixtures and to predict the properties that are not available but needed in the simulation of an oil spill (Riazi, 2005). As an example, if one needs to calculate how much oil would be vaporized after a certain time, the diffusion coefficient of oil vapors in the air, vapor pressure of oil, density, and molecular weight are needed. Furthermore, through appropriate characterization methods, crude oil should be divided into a number of pseudocomponents with known specifications (Riazi and Al-Enezi, 1999, 2002). Methods of estimation of critical data, density, vapor pressure, diffusion coefficient, viscosity, surface tension and solubility for pure hydrocarbons, petroleum products, and crude oils are discussed in this chapter.

The sources of oil spills, their occurrences, and causes are discussed in Chapter 3. A summary of information about the ten largest oil spills in the world is provided in Table 1.1. As can be seen from this table, the major causes of oil spill occurrence

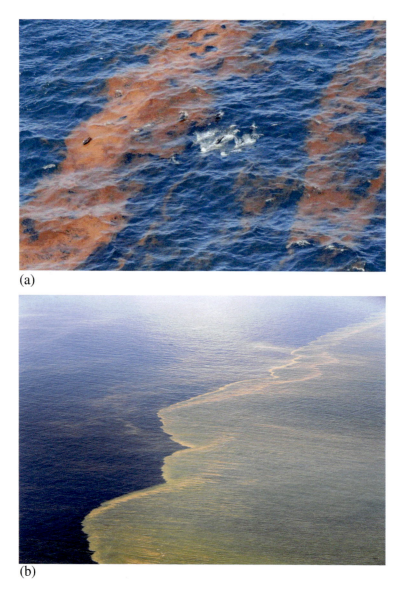

FIGURE 1.4 (a) Arial view of oil spill at the Gulf of Mexico after Deepwater Horizon disaster (NOAA photo). (b) Oil from the Deepwater Horizon oil spill approaches the coast of Mobile, Ala., May 6, 2010 (photo by United States Navy).

are well blowout during offshore activity, war, human error, and tanker collision as well as natural disasters such as storms. Other causes could be sabotage, operational issues, poor maintenance, and corrosion. Further discussion on these issues and examples for each case are given in Chapter 3. The chapter begins with review of oil and gas reserves in the world and their lifetime. Data on production and consumption

Introduction

TABLE 1.1
Size and Cause of Ten Major Oil Spills in the World [CNN News, 2019]

Order	Name	Date	Location	Cause	Approximate Amount of Oil Released	
					Million gallons	Million barrels
1	Gulf War	January 1991	Persian Gulf, Middle East	Iraq–Kuwait war	252–336	5–8
2	Deepwater Horizon	April–July 2010	Gulf of Mexico, US	Wellhead blowout	165–210	4–5
3	Ixtoc I	June 1979– March 1980	Bay of Campeche, Gulf of Mexico	Exploratory well blowout	140–150	3.3–3.5
4	Fergana Valley	March 1992	Uzbekistan	Oil well blowout	>88	>2
5	Nowruz Oil Field	February– September 1983	Persian Gulf/ Iran	Iraq–Iran war	>80	>1.9
6	Castillo de Bellver	August 1983	Cape Town, South Africa	Tanker catching fire	>78	>1.8
7	Amoco Cadiz	March 1978	Portsall, France	Tanker runs aground	>68	>1.6
8	Odyssey Tanker	November 1988	Newfoundland, Canada	Storm breaking tanker	>43	>1
9 (two spills)	Atlantic Empress and Aegean Captain	July 1979 and August 1979	Trinidad and Tobago/ Barbados	Two tankers collide while being towed away	>42.7 and >41.5	~1 and ~1
10	Production Well, D-103	August 1980	Tripoli, Libya	Well blowout	~42	~1

rates in different regions and trade movement maps for both oil and gas are presented for the year 2018 based on the BP statistical report in 2019. Natural gas usually is transported in the form of LNG, especially for long distances. Modes of transportation such as marine tankers and pipelines are discussed with data on the transportation routes and volume. The occurrence and causes of oil spills due to failure or accidents in offshore production wells and platforms, pipelines and shipping, oil tanker wrecks, onshore storage and pipelines, refining, and oil and gas stations near the sea are discussed. Finally some data on the amount and volume of oil versus the cause of oil spill are presented.

On the basis of offshore production-related oil spills, the BP or Gulf of Mexico (GOM) oil spill was the biggest oil spill in history, killing 11 people. The accident occurred on April 20, 2010, in the Gulf of Mexico at a production well (Macondo) about 70 km off the coast of Louisiana. The oil flow to water which initially was at a rate of 1,000–5,000 bbl/d increased gradually to about 57,000 bbl/d by June 2010. The GOM oil spill created a major environmental and economic catastrophe for both the US and the operating company BP (Riazi, 2016). Some 37% of the GOM area was closed for fishing activity, and the spill reached the shores of four states: Louisiana, Mississippi, Alabama, and Florida. It killed hundreds of birds and sea animals, closing many businesses in the affected areas. The damaged Macondo well was finally capped on July 15, 2010, after 87 days. During this period, according to US government officials, 4.9 million barrels of oil flowed into the seawater. As the event was fully covered by the media for over a year and it is the most recent major accident, more information on this oil spill is available. The cleanup operations took four years, costing BP tens of billions of dollars. It is for these reasons that Chapter 4 is devoted largely to this accident, and it revisits the BP-GOM oil spill as it happened. In this chapter meteorological data and flow rates as well as the composition and properties of the oil are given for simulation purposes. In the last part of the chapter, the trajectory of the spill as it was moving westward, as generated from the National Oceanic and Atmospheric Administration (NOAA) and other research institutes, from the time of the accident until the end of July 2010 is presented. In fact, April 20, 2020, was the tenth anniversary of this accident and one of the reasons for launching this book project. Chapter 4 begins with a brief history of oil spills around the world and reviews some major oil spills including the 1991 oil spill in the Persian Gulf as a result of Iraq's invasion of Kuwait. The Persian Gulf oil spill was the largest oil spill in human history at that time, releasing between 6 and 8 million barrels of oil into sea. For this reason, this event is also reviewed along with some other major accidents in Chapter 4.

As the BP oil spill was the greatest oil spill in US history and the world, with enormous impacts on the environment, health, and the economy, these issues are reviewed in Chapter 5. Figure 1.5 shows a brown pelican covered in oil from the GOM oil spill being washed at a cleaning center in Alabama. Figure 1.6 shows oil from the BP accident reaching marsh wetlands in Louisiana. Figure 1.7 shows how the oil from the seawater surface reflects the sunlight and thin layers are formed. Chapter 5 the effects of compounds from oil dissolved in water or in the air on the people in the surrounding areas are discussed. As the source of the leak was at the bottom of the sea, some light components from the oil can be dissolved in water while the oil is traveling the 5,000 feet from the leak source to the sea surface. As soon as the accident happened, BP operators began trying various methods never tested before in deep water to stop the flow of oil. Many methods, including using caps of different size, were tried until July 15 when the leak was sealed, and in September the Macondo well was permanently sealed off by cement through a method known as "static kill" as discussed in Chapter 5. In this chapter the political economic impacts of the Deepwater Horizon accident are also reviewed. The US president called the incident a national tragedy with impacts as great as 9/11

Introduction   7

FIGURE 1.5 Female brown pelican being rinsed following extensive cleaning at the Theodore Oiled Bird Rehabilitation Center in Alabama (photo by Tom MacKenzie, USFWS, May 26, 2010). USFWS: U.S. Fish & Wildlife Service.

FIGURE 1.6 Thick oil from the BP Deepwater Horizon oil spill floats on the surface of the water and coats the marsh wetlands in Bay Jimmy, Louisiana, June 11, 2010 (*The Atlantic*, July 2015).

and traveled four times to the region to monitor the situation. He also disused the matter with the UK prime minister and made a national speech asking BP to cover all the costs including fines and damages to the people in the region. The shares of BP dropped more than 50% from the time of the accident, and its CEO had to resign as a result of the disaster. US lawmakers grilled BP's top managers and made new regulations regarding offshore production activity in the Gulf of Mexico.

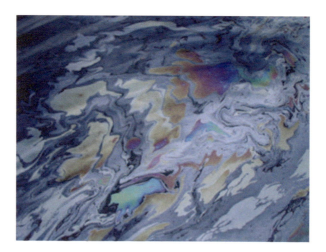

FIGURE 1.7 When medium and light oils spread unhindered, very thin films eventually form. These appear as an iridescent (rainbow) and silver sheen which dissipates rapidly (ITOPF, 2008).

Efforts to clean up the sea of oil usually begin immediately following an oil spill accident. Cleanup methods are complex and challenging and may include physical/mechanical, chemical, and biological techniques which are discussed in Chapter 6. The natural behavior of an oil spill is shown in Figure 1.8. Natural processes play an important role in the fate of an oil spill, but they are slow and time-consuming. Some of the equipment needed to accelerate cleanup operations are booms, skimmers, pumps, storage, dispersants and spray systems, response vessels, absorbents, and other spill-response equipment such as aircrafts and boats. The physical methods may include gravity separation, the use of booms and skimmers, flotation, and filtration while chemical methods include the use of chemicals such as inorganic and organic sorbents, dispersants, demulsifiers, and biosurfactants as well as in-situ burning, coagulation, and flocculation. Absorbent materials are those chemicals such as oleophilic material that have the capability of attracting oil, and then the oil and absorbent are removed together (Riazi, 2010). Some of these chemicals have their own environmental impacts, and for this reason bioremediation may be used as an alternative method to reduce environmental risk.

The bioremediation of oil involves the use of bioemulsifiers or biosurfactants as shown in Figure 1.9. Bioemulsifiers disperse the oil slick into smaller droplets and boost oil biodegradation by microbes, while the role of biosurfactants is to increase the solubility and bioavailability of hydrocarbons (Doshi et al., 2018). Some sorbent materials are capable of absorbing 15 g of oil per gram of sorbent. Generally, dispersants are based on hydrocarbon solvents with 15–25% dispersants and are used with a dosage of one to three dispersant to oil volume ratio. Another group of dispersants is used at a much lower dose. The solvent base is alcohol or glycol (oxygenated solvent) with a higher concentration of surfactant. These dispersants are used at a dose of 1 volume surfactant to 10–30 volume oil spill (Riazi, 2010). Some other

Introduction

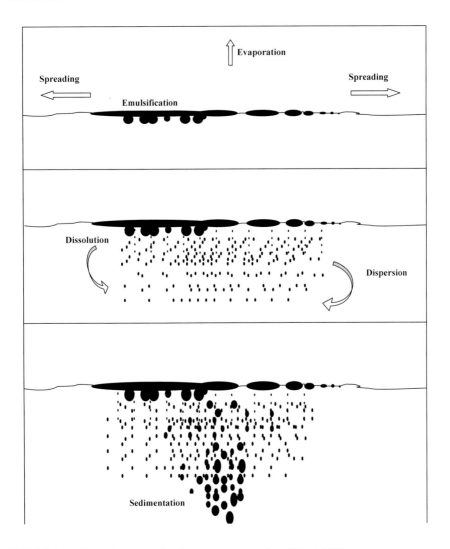

FIGURE 1.8 Behavior of an oil spill on seawater surface (Riazi, 2016).

dispersants are used to solidify oil from liquid to solids. These dispersants are polymer-based and usually used for inland oil spills (ITOPF, 2008). The Environmental Protection Agency (EPA) ordered BP to use a less toxic chemical oil dispersant to break up the oil in the Gulf of Mexico.

Spreading is the horizontal expansion of an oil slick due to gravity, inertia, viscous, interfacial tension, and turbulent diffusion. Spreading is a physical process in which oil spreads rapidly over a large area and breaks up in windrows which are long narrow slicks with the same direction as the wind. The rate of spreading very much depends on the oil density, viscosity and interfacial tension as well as wind and water speeds. A simple and approximate method was proposed by Lehr et al. (1984) to

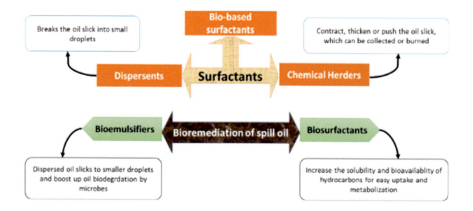

FIGURE 1.9 Chemical treatment of oil spill using surfactants and bioremediation method (Doshi et al., 2018).

predict the non-symmetrical expansion of oil slicks. The method predicts the length of the minor and major axes from water and oil densities, initial volume of spill, wind speed, and time. For an oil spill of 1 ton, after 10 minutes, the oil can disperse over a radius of 50 m, forming a slick 10 mm thick. The slick thickness decreases to less than 1 mm as oil continues to spread with an area of 12 km². As discussed by Riazi and Roomi (2008) and based on data published by Leber (1989) and Galt et al. (1991) for the case of the Exxon Valdez oil spill, the oil expanded 500 miles within 60 days of its release as shown in Chapter 7.

Skimming is a mechanical method in which oil from the water surface is removed by skimmers. Skimmers can recover oil at various rates, depending on oil slick thickness. For example, for slick of 1 mm thickness the recovery rate is about 35 tons/hr while for 6 mm thickness film the rate of oil removal is about 70 tons or 420 bbl/hr (Riazi, 2010). In-situ burning refers to burning oil in its place over the seawater surface. The largest in-situ burning occurred in Kuwait in 1991 which burned more than one billion barrels of oil over a nine-month period. Although in-situ burning can reduce oil disposal following a recovery, it produces a large amount of smoke with a negative impact on the environment (Allen and Ferek, 1993). The use of booms to control oil and to burn it at the sea surface for the case of the BP-GOM oil spill is shown in Figure 1.10 as was reported by ABC News (2010).

An important aspect of predicting oil spill behavior and choosing the most effective and appropriate response method is modeling and simulation. Figure 1.11 shows the predicted trajectory of the BP oil spill after seven weeks from the time of the accident and how it is spreading westward. The fate of an oil spill is determined by major dynamic processes that may occur once the oil is floating on the seawater surface. The behavior of oil spills largely depends on the type and nature of the oil as well as the environment in which it is spilled, such as air temperature, water temperature, water and wind speeds, and wave conditions. In general, the dynamic processes that marine-type oil spills may go through include spreading, evaporation, dissolution, dispersion, emulsification, sedimentation, and degradation processes as

Introduction

FIGURE 1.10 Smoke rises from a controlled burn in the Gulf of Mexico on May 19, 2010 (Wikinews, May 25, 2010).

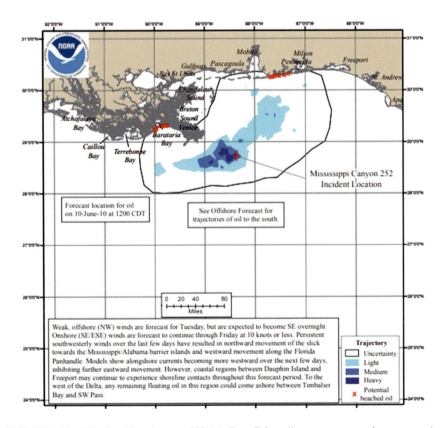

FIGURE 1.11 BP oil spill trajectory (NOAA, June 7, http://response.restoration.noaa.gov).

shown in Figure 1.8 (Riazi, 2010). NOAA predicted that about 17% of the oil was directly recovered from the damaged Deepwater Horizon well, while about 5% was burned at the surface and about 3% skimmed from the surface, about 5% chemically dispersed, 20% naturally dispersed, 25% evaporated, and as much as 25% remained as residual on the top or bottom of the sea as it was difficult to remove. Due to the importance of modeling and simulation methods to predict the behavior of oil spills, two chapters of the book are devoted to this topic.

In Chapter 7 simple methods are presented for quick estimates and approximate simulations of three dynamic processes that an oil spill will go through once released into the sea. These three processes are evaporation, dissolution, and sedimentation. The amount of dissolution is small because of the low solubility of hydrocarbons in water, but from a toxicological point of view it is very important to determine the amount of oil dissolved in water. In addition, the type of hydrocarbons dissolved is also important to the degree of toxicity. For example, mono-aromatics are the most toxic type of hydrocarbons, and for this reason it is important to determine the concentration of hydrocarbon type versus time (Riazi and Roomi, 2005, 2006, 2008). The amount of evaporation of oil spill largely depends on the concentration of more volatile compounds (light hydrocarbons), while the amount of sedimentation depends on the distribution of heavy components which are heavier than water. Two models are presented in Chapter 7. The proposed model is based on laboratory data, and it is assumed that a certain and fixed volume of oil is released on the water surface and the amount of oil vaporized, dissolved, or sedimented into the bottom of the sea can be estimated. This is the case when oil spills are caused by a tanker collision and the exact amount of oil released into the water is known. The area of oil spill and its volume on the water surface versus time due to weathering can be estimated. In the last part, the model is extended to cases when there is a variable flow of oil onto the water surface. The model can only be applied to water at the surface and not in water columns. Its application to the Deepwater Horizon (BP) oil spill just for the amount of oil reaching the surface of the water is demonstrated. Daily variation of temperature and wind speed are considered in the numerical simulation of oil leaked from offshore damaged wells. As mentioned earlier, these models are approximate, and they can be used for an initial and quick estimate of the rate of oil spill disappearance to evaluate the most effective response technique. Many other processes such as spreading, emulsification, dispersion, biodegradation, and oxidation are not included in the simplified model presented in Chapter 7.

More comprehensive models include the rate of spreading, horizontal movement of oil slick, and vertical distribution of oil droplets. Commercial software such as OILMAP, TRANSAS, OILFOW2D, OSCAR, or ANSYS is based on simulating the processes such as spreading, horizontal turbulent diffusion, advection, vertical dispersion, emulsification, and evaporation and is capable of predicting the horizontal movement of oil slicks as discussed in an article review by Kastrounis (2018). In these simulations, computational fluid dynamics (CFD) and the application of the finite volume method to solve the Navier–Stokes (continuity and momentum) governing equations have been applied (Guo and Wang, 2009). Some of these advanced simulators are presented and discussed in Chapter 8. One publicly available model

Introduction 13

was developed by NOAA and is called the General NOAA Operational Modeling Environment (GNOME). It is a tool that the Office of Response and Restoration's Emergency Response Division uses to predict the possible route or trajectory of an oil spill. The model predicts how wind, currents, and other processes might move and spread oil on the water. It also predicts how oil is expected to change physically and chemically, which is known as weathering, during the time that it remains on the water surface (GNOME, 2002). The latest version of GNOME was published in November 2017 by NOAA.

Chapter 8 starts with the latest developments in 3D oil spill modeling, that is, the modeling of deep-water oil well blowout and the subsequent creation of an oil plume. Oil jets created from subsea blowouts are challenging due to the multiple physical processes and forces involved. The loss of pressure leads to expansion and possibly to the formation of gas that accelerates the fluid further. A short distance from the oil exit, the fluid changes from jet-like to plume-like. The resulting plume contains liquid droplets, gas bubbles, entrained water, and potentially gas hydrates. An example of plume simulation from OSCAR Deep Blow is shown in Figure 1.12 where the model tracks the plume as a multi-phase volume, including droplets, bubbles, dissolved compounds, entrained seawater, and gas hydrates. Two types of models are discussed: one for near-field and one for far-field oil spills. Characteristics and specifications of models such as BLOSOM, MIKE, MOHID, OILMAP, OSCAR, and TAMOC are discussed in detail.

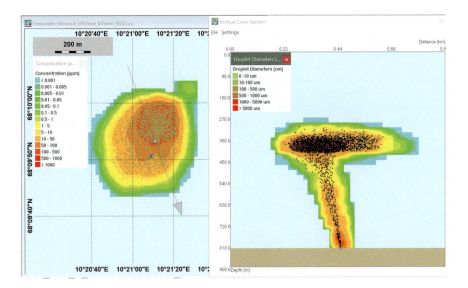

FIGURE 1.12 Example of plume simulation from OSCAR DeepBlow. On the left, the color scale indicates different concentrations of dissolved compounds. On the right, one can see the plume reaching the trap height, forming an intrusion layer (simulation animation by Konstantinos Kotzakoulakis and available at https://youtu.be/KHQUztxgG50). Taken from Chapter 8.

Open-source models such as TAMOC, OpenDrift, and GNOME are developed by the academic community and public institutes, and are freely offered for use and customization. This ability makes them ideal for researchers wanting to experiment with new parametrizations, and students wanting to learn the inner workings of a simulator. In contrast, using commercial models such as MIKE and OSCAR involves a cost but provides state-of-the-art parametrizations, a detailed graphical interface, technical support, and modeling services. These characteristics make them the better choice for industry clients and decision-making authorities as discussed in Chapter 8.

The focus of Chapter 9 is on the economic and financial aspects of offshore oil spills. The local economic and financial impacts include cleanup costs, natural resource damage, socioeconomic losses, and other related costs. The global impacts of oil spills include oil price, financial markets, and energy industry network. In this chapter some cost estimation models are presented for cleanup operation with some case studies in the UK and Finland. In one model, the costs are divided into on-water and shoreline costs based on some case studies. In shoreline cleanup costs, a method is presented for the estimation of the lost service value of natural resources after oil impacts. The oil pollution will also cause losses to various socioeconomic aspects, e.g. fishery and tourism, by reducing their profits. At the end of the chapter some methods of damage estimation and compensation for some major oil spill incidents caused by tanker accidents are presented.

REFERENCES

ABC News. 2010. *Taken from the following link on May 23, 2010 at 9:00 am ET.* https://abcnews.go.com/

Allen, A.A. and R.J. Ferek. 1993. Advantages and disadvantages of burning spilled oil. In *Proceedings of the 1993 international oil spill conference.* American Petroleum Institute, Washington, DC, pp. 765–772.

BBC News. 2019. June 15. https://www.bbc.com/news

CNN News. 2019. *Oil spills fast facts.* CNN Library, Updated April 22, 2019. https://www.cnn.com/2013/07/13/world/oil-spills-fast-facts/index.html

Deepwater Horizon Report. 2011. *On scene coordinator report deep water horizon oil spill, National Response Team (NRT).* September. https://homeport.uscg.mil/Lists/Content/Attachments/119/DeepwaterHorizonReport%20-31Aug2011%20-CD_2.pdf

Doshi, B., M. Sillanpaa and S. Kalliola. 2018. A review of bio-based materials for oil spill treatment. *Water Research*, 135, May 15, pp. 262–277.

Galt, J.A., W.J. Lehr and D.L. Payton. 1991. Fate and transport of Exxon Valdez oil spill. *Environmental Science & Technology*, 25(2), p. 202.

Guo, W.J. and Y.X. Wang. 2009. A numerical oil spill model based on a hybrid method. *Marine Pollution Bulletin*, 58(5), pp. 726–734.

Haven, M.T. 1991. *Science, taken from the following site on June 15, 2019.* https://sciencele.weebly.com/oil-spill.html

IEA Report. 2018. *Offshore energy outlook, International Energy Agency (IEA), Paris based intergovernmental organization, reported on May 4.* https://www.iea.org/weo/offshore/ (accessed on June 15, 2019).

ITOPF. 2008. *The international tanker owners Pollution Federation Ltd.* London. http://www.itopf.com/spill-response/clean-up-and-response/alternative-techniques/

Introduction

Kastrounis, N. 2018. Review of oil spill simulation, DEMSEE'18. In *13th international conference on deregulated electricity market issues in South Eastern Europe*. Nicosia, Cyprus.

Leber, P.A. 1989. *Environmental chemistry: A case study of the Exxon Valdez oil spill of 1989*. Franklin and Marshall College, Lancaster, PA. http://wulfenite.fandm.edu/exxon-valdez.htm.

Lehr, W., H. Cekirge, R. Fraga and M. Belen. 1984. Empirical studies of the spreading of oil spills. *Oil and Petrochemical Pollution*, 2, pp. 7–11.

MEO. 2019. *Middle east online*. July 25. https://meo.news/en/britain-escort-all-uk-vessels-through-hormuz-strait

NOAA. 2002. *GNOME (General NOAA Operational Modeling Environment) user's manual, National Oceanic and Atmospheric Administration (NOAA)*. January. https://response.restoration.noaa.gov/oil-and-chemical-spills/oil-spills/response-tools/gnome.html

Planete Energies. 2015. *Transporting oil by sea*. January 14. https://www.planete-energies.com/en/medias/close/transporting-oil-sea

Riazi, M.R. 2005. *Characterization and properties of petroleum fractions, ASTM manual 50*. ASTM International, Conshohocken, PA, p. 435.

Riazi, M.R. 2010. *Accidental oil spills and control*, Chapter 5 in the book entitled "Environmentally Conscious Fossil Energy Production", edited by Myer Kutz and A. Elkamel, John Wiley and Sons, New York.

Riazi, M.R. 2016. Modeling and predicting the rate of hydrocarbon vaporization from oil spills with continuous oil flow. *International Journal of Oil, Gas, and Coal Technology (IJOGCT)*, 11(1), pp. 93–105.

Riazi, M.R. and G. Al-Enezi. 1999. A mathematical model for the rate of oil spill disappearance from seawater for kuwaiti crude and its products. *Chemical Engineering Journal*, 73, pp. 161–172.

Riazi, M.R. and G. Al-Enezi. 2002. *A model for oil dissolution in seawater*. Presented at the IASTED international conference on modeling and simulation (MS 2002). Marina del Rey, California, U.S.A., May 13-15.

Riazi, M.R. and M. Edalat. 1996. Prediction of the rate of oil removal from seawater by evaporation and dissolution. *Journal of Petroleum Science and Engineering*, 16, pp. 291–300.

Riazi, M.R. and Y.A. Roomi. 2005. *A predictive model for the rate of dissolution of oil spill and its toxic components in sea water*. Presented at the 230th ACS annual meeting, division of petroleum health and safety, Washington, DC, August 28–September 1.

Riazi, M.R. and Y.A. Roomi. 2006. *Solubility of toxic compounds from petroleum spills into seawater*. The 20th annual European simulation and modelling conference, ESM'2006, Toulouse, France, October 23–25.

Riazi, M.R. and Y.A. Roomi. 2008. A model to predict rate of dissolution of toxic compounds into seawater from an oil spill. *International Journal of Toxicology*27 (5) (2008), pp. 379–386, September 1, pp. 379–386.

Russia Today. 2019. *News*. June 15. https://www.rt.com/news/

Villoria, C.M., A.E. Anselmi, S.A. Intevep and F.R. Garcia. 1991. An oil spill fate model. *Society of Petroleum Engineers*, SPE23371, pp. 445–454.

2 Characteristics and Properties of Crude Oil and Its Products

NOMENCLATURE

API	API gravity defined in Equation 2.2
$a,b...i$	Correlation constants in various equations
BI_{vis}	Blending index for viscosity of liquid hydrocarbons (see Equation 2.64), dimensionless
CH	Carbon-to-hydrogen weight ratio
D_{AB}	Binary (mutual) diffusion coefficient (diffusivity) of component A in B, cm^2/s
$D_{A\text{-}mix}$	Effective diffusion coefficient (diffusivity) of component A in a mixture, cm^2/s
d_T	Liquid density at temperature T and 1 atm, g/cm^3
d_{20}	Liquid density at 20°C and 1 atm, g/cm^3
$\hat{f}_i$	Fugacity of component i in a mixture defined by Equation 2.85, bar
f_i^L	Fugacity of pure liquid i at T and P, bar
GOR	Gas-to-oil ratio (scf/bbl)
I	Refractive index parameter defined in Equation 2.11
K_W	Watson (UOP) K factor defined by Equation 2.8
k_i	Henry's law constant defined by Equation 2.84, bar
$k_{gas,water}$	Henry's law constant of a gas in water (Equation 2.85), bar
M	Molecular weight, g/mol [kg/kmol]
n	Sodium D line refractive index of liquid at 20°C and 1 atm, dimensionless, defined in Equation 2.10
N_C	Carbon number (number of carbon atoms in a hydrocarbon molecule)
P_c	Critical pressure, bar
P^{vap}	Vapor (saturation) pressure, bar
P_r	Reduced pressure (= P/P_c), dimensionless
P_r^{vap}	Reduced vapor pressure at a given temperature (= P^{vap}/P_c), dimensionless
R	Universal gas constant, 8.314 J/mol·K
R_i	Refractivity intercept in Equation 2.9
SG	Specific gravity of liquid substance at 15.5°C (60°F) defined by Equation 2.1, dimensionless
SG_g	Specific gravity of gas substance at 15.5°C (60°F) defined by Equation 2.3, dimensionless
T_b	Boiling point, K

T_c	Critical temperature, K
T_F	Flash point, K
T_M	Melting (freezing point) point, K
V	Molar volume, cm³/gmol
V_c	Critical volume (molar), cm³/mol (or critical specific volume, cm³/g)
x_{wi}	Weight fraction of component i in a mixture (usually used for liquids), dimensionless
x_P, x_N, x_A	Fractions (i.e., mole) of paraffins, naphthenes, and aromatics in a petroleum fraction, dimensionless
y_i	Mole fraction of i in a mixture (usually used for gases), dimensionless
Z	Compressibility factor (= PV/RT), dimensionless
Z_c	Critical compressibility factor ($Z=P_cV_c/RT_c$), dimensionless

GREEK LETTERS

μ	Absolute (dynamic) viscosity, cp [mPa·s]; also used for dipole moment
ν	Kinematic viscosity defined by Equation 2.7, cSt [mm²/s]
$\nu_{39(100)}$	Kinematic viscosity of a liquid at 39°C (100°F), cSt (10⁻² cm²/s)
δ_i	Solubility parameter for i, (J/cm³)^{1/2} or (cal/cm³)^{1/2}
θ	A property of hydrocarbon such as: $M, T_c, P_c, V_c, I, d, T_b, \ldots$
ρ	Density at a given temperature and pressure, g/cm³
σ	Surface tension, dyn/cm [= mN/m]
σ_H	Surface tension of a hydrocarbon at a given temperature, dyn/cm
σ_{wo}	Interfacial tension of oil and water at a given temperature, dyn/cm
ω	Acentric factor defined by Equation 2.5, dimensionless
ξ	Viscosity parameter defined by Equation 2.72, (cp)⁻¹
γ_i	Activity coefficient of component i in liquid solution, dimensionless

SUBSCRIPTS

A	Aromatic
N	Naphthenic
P	Paraffinic
T	Value of a property at temperature T
r	Reduced property, dimensionless
°	A reference state for T and P
∞	Value of a property at $M \rightarrow \infty$
20	Value of a property at 20°C
39(100)	Value of kinematic viscosity at 39°C (100°F)
99(210)	Value of kinematic viscosity at 99°C (210°F)

ACRONYMS

API-TDB	American Petroleum Institute – Technical Data Book
EOS	Equation of state

Characteristics and Properties of Petroleum 19

IUPAC International Union of Pure and Applied Chemistry
GOR Gas-to-oil ratio

2.1 NATURE OF CRUDE OILS AND THEIR PRODUCTS

Petroleum is a complex mixture of hundreds of hydrocarbon compounds of different types which is stored in sedimentary rocks in the form of gas or liquid called reservoir fluid. The ratio of gas to oil in a reservoir fluid is called the gas-to-oil ratio (GOR) and is an important characteristic of reservoir fluids. Once a reservoir fluid is produced from a production well at a surface facility, it is brought to surface conditions (usually at 1 atm and 60°F) through multistage separators. Produced liquid is generally referred to as crude oil, and the gas is referred to as associate gas. The general composition of a reservoir fluid with produced liquid and gas from a petroleum reservoir in Kuwait is shown in Table 2.1 (ASTM MNL50). The C_{7+} fraction represents all compounds with carbon number ≥ 7.

As shown in this Table 2.1 a crude oil is mainly a mixture of hydrocarbons with some heteroatoms such as sulfur (S), nitrogen (N), oxygen (O), and metals. The

---

TABLE 2.1

Composition of a Reservoir Fluid and Flashed Gas and Produced Crude Oil from a Reservoir in the Middle East

No.	Component	Reservoir Fluid	Flashed Gas	Liquid Crude Oil
1	CO_2	1.85	2.91	0
2	N_2	0.07	0.11	0
3	H_2S	1.55	2.42	0
4	C_1	29.87	47	0
5	C_2	14.24	22.21	0.38
6	C_3	9.28	13.98	1.9
7	nC_4	4.02	5.35	2
8	iC_4	0.91	1.28	0.8
9	nC_5	2.15	1.97	3.9
10	iC_5	1.18	1.23	2.3
11	C_6	2.88	1.14	6.6
12	C_{7+}	32.00	0.4	82.12
Sp. gr. @ 60°F		0.715		0.824
MW		88.99	30.0	177
Temp, °F		241		90
Pressure, psia		2280		15
V/F ratio			0.6362	0.3632
GOR [scf/stb]		872		

Note: Numbers represent mol% of each component.

composition (wt%) of the elements depends on the source of the crude and varies within narrow limits as:

Carbon (C), 83.0–87.0%
Hydrogen (H), 10.0–14.0%
Nitrogen (N), 0.1–2.0%
Oxygen (O), 0.05–1.5%
Sulfur (S), 0.05–6.0%
Metals (nickel, vanadium, and copper), <1000 ppm (0.1%)

Generally in heavier oils, amounts of carbon and heteroatoms increase with increase in density (lower API gravity). Hydrocarbons found in petroleum oils are mainly from four families or groups: (1) paraffins, (2) olefines, (3) naphthenes, and (4) aromatics. Paraffins, olefins, and naphthenes are sometimes called *aliphatic* versus aromatic compounds. Naphthenes and aromatics are cyclic compounds, while paraffins and olefins are non-cyclic. The simplest hydrocarbon compound is methane from the paraffin group. Paraffins are divided into two groups of normal paraffin (also called n-alkanes) and iso-paraffins (or branched paraffins). The general closed formulae for these groups are: C_nH_{2n+2} (paraffins), C_nH_{2n} (olefins), C_nH_{2n} (naphthenes), and C_nH_{2n-6} (where $n \geq 6$) for aromatics. Structures of some sample compounds are shown below:

n-Butane (n-C_4H_{10} or simply shown as n-C4) is an example of an n-paraffin or n-alkane compound.

$$\begin{array}{c} H \quad H \quad H \quad H \\ | \quad\; | \quad\; | \quad\; | \\ H - C - C - C - C - H \\ | \quad\; | \quad\; | \quad\; | \\ H \quad H \quad H \quad H \end{array}$$

Iso-octane (2-methylheptane) is an example of a branched paraffinic compound (C_8H_{18}).

$$\begin{array}{c} CH_3 \\ | \\ CH_3\text{-}CH\text{-}CH_2\text{-}CH_2\text{-}CH_2\text{-}CH_2\text{-}CH_3 \end{array}$$

Compounds with one double bond (also called mono-olefins) or alkenes such as ethene also named ethylene ($CH_2{=}CH_2$) and propene or propylene ($CH_2{=}CH{-}CH_3$) are from olefinic compounds. Since olefins are generally unstable compounds, usually most petroleum products are olefin-free.

A few sample compounds from the naphthenic group are shown here:

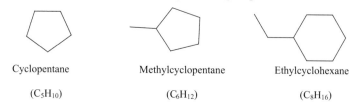

Cyclopentane (C_5H_{10})   Methylcyclopentane (C_6H_{12})   Ethylcyclohexane (C_8H_{16})

Characteristics and Properties of Petroleum

Some examples of aromatic compounds are:

(C_6H_6)	(C_7H_8)	(C_8H_{10})	($C_{10}H_8$)
Benzene	Toluene	O-xylene	Naphthalene
	(Methylbenzene)	(1,2-Dimethylbenzene)	

Compounds with methyl groups have an attachment of CH_3 and the cycles with double bonds are attached by CH. Heavier oils have higher amounts of polynuclear aromatic hydrocarbons (PAHs). Sulfur and nitrogen are usually present in PAHs such as:

Dibenzothiophene   Benzocarbazole ($C_{16}H_{11}N$)

Sulfur also may exist in non-cyclic compounds such as mercaptans (R–S–H) and sulfides (R–S–R′) where R and R′ are alkyl groups. Asphaltene is an aromatic compound (PAH) that may exist in the residues of some heavy oils and has a molecular weight of 1,000–5,000 g/mol and density of 1.1–1.2 g/ml. A sample molecule of asphaltene is shown in Figure 2.1 (Speight, 1998).

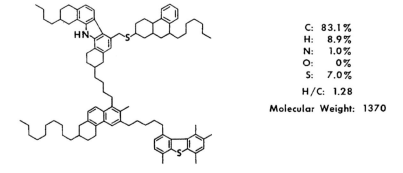

C: 83.1%
H: 8.9%
N: 1.0%
O: 0%
S: 7.0%
H/C: 1.28
Molecular Weight: 1370

FIGURE 2.1 An example of an asphaltene molecule.

TABLE 2.2

Some Petroleum Fractions Produced from Distillation Columns

Petroleum Fraction	Approximate Hydrocarbon Range	Approximate Boiling Range	
		°C	°F
Light gases	C_2–C_4	−90 to −1	−130 to 30
Gasoline (light and heavy)	C_4–C_{10}	−1 to 200	30 to 390
Naphthas (light and heavy)	C_4–C_{11}	−1 to 205	30 to 400
Jet fuel	C_9–C_{14}	150 to 255	300 to 490
Kerosene	C_{11}–C_{14}	205 to 255	400 to 490
Diesel fuel	C_{11}–C_{16}	205 to 290	400 to 550
Light gas oil	C_{14}–C_{18}	255 to 315	490 to 600
Heavy gas oil	C_{18}–C_{28}	315 to 425	600 to 800
Wax	C_{18}–C_{36}	315 to 500	600 to 930
Lubricating oil	>C_{25}	>400	>750
Vacuum gas oil	C_{28}–C_{55}	425 to 600	800–1100
Residuum	>C_{55}	>600	>1100

Crude oil is fed into an oil refinery where it first goes through atmospheric and then vacuum distillation columns. Some products of crude oil distillation with their carbon number and boiling ranges are given in Table 2.2 (Riazi, 2007).

Based on the API gravity and viscosity of oil, a general classification of crude oil is given in Table 2.3. According to this classification, the term heavy oil refers to oils with API degrees between 10 and 22, while oils with API gravity of greater than 22 are referred to as light oil. If the API gravity is less than 10 but the oil under reservoir conditions is mobile it is referred to as extra heavy oil, and if immobile (viscosity greater than 10,000 cSt) it is referred to as bitumen or oil sand.

TABLE 2.3

Classification of Oils and Definition of Heavy Oil, Extra Heavy Oil and Bitumen

Type of Oil	API Gravity	Viscosity in the Reservoir	Definition of Oil
Conventional Oil	>45°		Condensate
	22° to 45°		Medium–light crude
Non-Conventional Oil	10° to 22°	Between 100 and 10000 cSt at reservoir conditions	Heavy crude
	<10°	<10000 cSt Mobile at reservoir conditions	Extra heavy crude
	<10°	>10000 cSt Immobile at reservoir conditions	Bitumen

Characteristics and Properties of Petroleum 23

2.2 BASIC PROPERTIES OF PURE HYDROCARBONS

In this section, some basic properties that are needed in the characterization of petroleum fractions or crude oil as well as those properties needed to estimate thermophysical properties are defined.

Molecular weight or molar mass, shown by M, is a number that represents the mass of one mole of the substance. M has the unit of kg/kmol in the SI and the unit of lb/lbmol in the English unit system. For pure compounds, the molecular weight is the sum of atomic weights of all elements in the molecule. The atomic weights of some elements in petroleum are: $C = 12.011$, $H = 1.008$, $S = 32.065$, $O = 16.0$, $N = 14.01$ as given by the IUPAC standard (Coplen, 2001). Molecular weight is an important characterization parameter for pure hydrocarbons and petroleum fractions.

The **boiling point** of a pure compound at a given pressure is the temperature at which vapor and liquid exist together at equilibrium. However, at 1 atmosphere the boiling point known as the normal boiling point is shown by T_b, and this is another important characteristic of petroleum fractions and pure compounds. Since petroleum products are mixtures of hydrocarbon compounds, they have a range of boiling points as shown in Table 2.2, and the mid-boiling point or an average boiling point represents the mixture boiling point. For heavy hydrocarbons, boiling points at lower pressures are reported, as, due to thermal cracking of hydrocarbon bonds and decomposition of the compounds, the normal boiling point cannot be measured.

Density is defined as mass per unit volume, and it is shown by "d" or "ρ" and has the unit of kg/m³. Density is a function of both temperature and pressure; however, the effect of pressure on the density of liquids is much smaller than the effect of pressure on the density of gases. Usually the liquid density of oil at 1 atm and 20°C (shown by d_{20}) is measured in laboratories and reported as a characterization parameter of oil. The **specific volume** is the reciprocal of density and represents volume of unit mass ($v = 1/d$). The molar volume is the volume of one mole of substance and has the unit of mol/m³ or lb_{mol}/ft^3 ($V = M/d$). Liquid densities are usually reported in terms of specific gravity (SG) which is the ratio of density to the density of water at the same temperature. In the petroleum industry the standard conditions are 60°F (15.5°C) and 1 atmosphere where the density of water is 0.999 g/cm³ and for a liquid petroleum fraction the **specific gravity** is defined as:

$$SG \ (60\ °F/60°F) = \frac{\text{density of liquid at 60 °F in g/cm}^3}{0.999 \text{ g/cm}^3} \tag{2.1}$$

The value of specific gravity is nearly the same as the density of liquid at 15.5°C (289 K) in g/cm³. Another characterization parameter defined by the American Petroleum Institute (API) is the **API gravity** which is defined as (API-TDB, 1997):

$$\text{API Gravity} = \frac{141.5}{SG \ (@60\ °F)} - 131.5 \tag{2.2}$$

Hydrocarbons with lower specific gravities have higher API gravity. Aromatic hydrocarbons have higher specific gravity (lower API gravity) than paraffinic hydrocarbons.

For gases, the specific gravity (SG_g) is defined as the ratio of the density of the gas to the density of air at standard conditions which is equivalent to the ratio of the molecular weight of the gas (M_g) to the molecular weight of air (28.97):

$$SG_g = \frac{M_g}{28.97} \tag{2.3}$$

The properties of petroleum fluids usually are estimated through equations of state or generalized correlations which require critical constants and acentric factor as the input parameters. The critical point is a point on the PVT diagram where the liquid and vapor phases are identical and indistinguishable. The temperature, pressure, and volume at the critical point are called **critical temperature** (T_c), **critical pressure** (P_c), and **critical volume** (V_c) which are specific characteristics of a substance. For mixtures and petroleum fractions, as true critical properties are not measurable, they are referred to as *pseudocritical properties* and should be estimated. The **critical compressibility factor** (Z_c) is a dimensionless parameter defined as

$$Z_c = \frac{P_c V_c}{R T_c} \tag{2.4}$$

where R is the universal gas constant (= 8.314 J/mol·K). **Acentric factor** is a dimensionless parameter shown by ω and is defined as:

$$\omega = -\log_{10}\left(P_r^{vap}\right) - 1.0 \tag{2.5}$$

where P_r^{vap} is reduced vapor pressure (P^{vap}/P_c) at $T = 0.7 T_c$ (reduced temperature of 0.7).

One important property that directly affects the rate of vaporization of oil to the environment is **vapor pressure** (P^{vap}) which is defined as the force exerted per unit area of walls by the vaporized portion of the liquid in a closed container. Vapor pressure or saturation pressure (P^{sat}) can also be defined as the pressure at which vapor and liquid are in equilibrium at a given temperature. The vapor pressure of a compound is a measure of its volatility, and heavier oils have lower vapor pressures. At normal boiling point the vapor pressure of any substance is 1 atmosphere. Vapor pressure increases with temperature, and the maximum vapor pressure is the critical pressure at the critical temperature. When a liquid hydrocarbon is open to the atmosphere at a given temperature at which the vapor pressure of liquid is P^{vap}, the volume% of the hydrocarbon vapors in the air is:

$$Vol.\% = 100 \times \left(\frac{P^{vap}}{P_a}\right) \tag{2.6}$$

Characteristics and Properties of Petroleum

where P_a is the atmospheric pressure. At sea level where $P_a = 1$ atm, the calculation of vol% of hydrocarbon vapor in the air from Equation 2.6 is simply $100P^{vap}$ when P^{vap} is in atm.

Another measurable property for petroleum products is the kinematic viscosity (ν) which is defined as the ratio of viscosity, μ, to density ρ at the same temperature in the following form:

$$\nu = \frac{\mu}{\rho} \tag{2.7}$$

Viscosity, μ, is a molecular property defined from Newton's law of viscosity and $1/\mu$ is called fluidity. Liquids with lower viscosity flow more easily and have less resistance to flow. Viscosity is usually measured in cP (centipoise) which is equivalent to mPa (milli Pascal). Kinematic viscosity is expressed in cSt (centistokes) and is usually measured and reported at two reference temperatures of 38°C (100°F) and 99°C (210°F). For a pure hydrocarbon, the **freezing point** is a temperature at which liquid solidifies at 1 atmosphere of pressure. For petroleum mixtures there is a range of freezing points, and it is related to the wax appearance temperature of the oil. Heavier oils with higher wax contents have higher freezing points.

Another important property of fuel oil is the **flash point** (T_F), which is defined as the minimum temperature at which the vapor pressure of oil is sufficient to produce vapor needed for the spontaneous ignition of fuel with the air in the presence of an external source (spark or flame). The flash point is directly related to the boiling point, and hydrocarbons with higher boiling points have higher flash points.

There are some defined characterization parameters that are useful in determining the type of a hydrocarbon compound. The **UOP characterization factor** (K_W) is one of the oldest characterization factors, defined in the 1930s by Watson et al. (1935) as:

$$K_W = \frac{\left(1.8T_b\right)^{1/3}}{SG} \tag{2.8}$$

where T_b is the normal boiling point in degrees Kelvin and SG is specific gravity at 15.5°C. This factor was defined to classify types of hydrocarbons. Aromatics have low K_W values while paraffins have high values and naphthenes have K_W values between these two. Another useful characterization parameter is the *refractivity intercept*, R_i, defined as:

$$R_i = n - \frac{d}{2} \tag{2.9}$$

where n and d are the refractive index and density (in g/cm³) of liquid hydrocarbon at 20°C and 1 atm. R_i is high for aromatics and low for naphthenic compounds, while paraffins have R_i values between values for these two groups. The **carbon-to-hydrogen weight ratio**, *CH*, is defined as the ratio of total weight of carbon atoms to the total weight of hydrogen in a compound or a mixture. The *CH* ratio increases from paraffins to naphthenes and to aromatics as the number of hydrogen atoms decreases.

Another basic property which is easily measurable in laboratories is the refractive index or refractivity (n) which is defined as the ratio of speed of light in a vacuum to the speed of light in the substance which is a function of temperature and pressure.

The **Refractive index** (n) is a property that can be easily and accurately measured in a laboratory and is defined as:

$$n = \frac{\text{velocity of light in the vacuum}}{\text{velocity of light in the substance}} \qquad (2.10)$$

Usually the refractive index of hydrocarbons is measured by the sodium D-line at 20°C and 1 atmosphere. The instrument to measure the refractive index is a refractometer and it can be used easily to determine the refractivity of liquids with good accuracy. This parameter is correlated through parameter I defined as:

$$I = \frac{n^2 - 1}{n^2 + 2} \qquad (2.11)$$

where I represents the ratio of actual molar volume to apparent molar volume and can be used as a characterization parameter for petroleum fractions.

$$I = \frac{R_m}{V} = \frac{\text{actual molar volume of molecules}}{\text{apparent molar volume of molecules}} \qquad (2.12)$$

The properties of some pure hydrocarbons from different groups, as reported in ASTM Manual 50 (Riazi, 2007), are given in Table 2.4. For four homologous groups of n-alkanes, n-alkylcycopentanes, n-alkylcyclohexanes, and n-alkylbenzenes the following correlation may be used to estimate basic properties in terms of molecular weight, M (Riazi and Al-Sahhaf, 1995).

$$\ln(\theta_\infty - \theta) = a - bM^c \qquad (2.13)$$

where θ is a property such as T_c and θ_∞ represents the value of the same property at $M \longrightarrow \infty$. For this equation, the constants for the freezing point, T_M, boiling point, T_b, specific gravity, SG, liquid density at 20°C, d_{20}, refractive index parameter, I, reduced boiling point, T_{br} (T_b/T_c), critical pressure, P_c, critical density, d_c, acentric factor, ω, and surface tension, σ, are given in Table 2.5 (Riazi, 2007).

2.3 BASIC PROPERTIES OF CRUDE OIL AND ITS PRODUCTS

The simplest method to estimate the properties of a narrow range petroleum fractions or products is to assume each mixture behaves as a pseudocompound similar to a pure hydrocarbon. The basic properties that should be estimated for such fractions or products in order to estimate various thermophysical properties are composition, critical properties, acentric factor, molecular weight, boiling point, density,

Characteristics and Properties of Petroleum

TABLE 2.4
Basic Physical Properties of Some Selected Compounds

No.	Compound	Formula	N_C	M	T_M °C	T_b °C	SG@60 °F	d_{20} g/cm³	n20	T_C °C	P_C bar	V_C cm³/mol	Z_C	ω
Paraffins														
1	Methane	CH_4	1	16.0	−182.5	−161.5	0.2999	-----	-----	−82.59	45.99	98.65	0.2864	0.0115
2	Ethane	C_2H_6	2	30.1	−182.8	−88.6	0.3554	0.3386	-----	32.17	48.72	145.48	0.2792	0.0995
3	Propane	C_3H_8	3	44.1	−187.7	−42.0	0.5063	0.4989	-----	96.68	42.48	200.14	0.2765	0.1523
4	n-Butane	C_4H_{10}	4	58.1	−138.3	−0.5	0.5849	0.5791	1.3326	151.97	37.96	255.09	0.2740	0.2002
5	n-Pentane	C_5H_{12}	5	72.2	−129.7	36.1	0.6317	0.6260	1.3575	196.55	33.70	313.05	0.2702	0.2515
6	n-Hexane	C_6H_{14}	6	86.2	−95.3	68.7	0.6651	0.6605	1.3749	234.45	30.25	371.22	0.2661	0.3013
7	n-Heptane	C_7H_{16}	7	100.2	−90.6	98.4	0.6902	0.6857	1.3876	267.05	27.40	427.88	0.2610	0.4395
8	n-Octane	C_8H_{18}	8	114.2	−56.8	125.7	0.7073	0.7031	1.3974	295.55	24.90	486.35	0.2561	0.3996
9	n-Nonane	C_9H_{20}	9	128.3	−53.5	150.8	0.7220	0.7180	1.4054	321.45	22.90	543.67	0.2519	0.4435
10	n-Decane	$C_{10}H_{22}$	10	142.3	−29.6	174.2	0.7342	0.7302	1.4119	344.55	21.10	599.58	0.2463	0.4923
11	n-Undecane	$C_{11}H_{24}$	11	156.3	−25.6	195.9	0.7439	0.7400	1.4151	365.85	19.50	658.69	0.2418	0.5303
12	n-Dodecane	$C_{12}H_{26}$	12	170.3	−9.6	216.3	0.7524	0.7485	1.4195	384.85	18.20	715.67	0.2381	0.5764
13	n-Tridecane	$C13H_{28}$	13	184.4	−5.4	235.5	0.7611	0.7571	1.4235	401.85	16.80	774.60	0.2319	0.6174
14	n-Tetradecane	$C_{14}H_{30}$	14	198.4	5.9	253.6	0.7665	0.7627	1.4269	419.85	15.70	829.82	0.2261	0.6430
15	n-Pentadecane	$C_{15}H_{32}$	15	212.4	9.9	270.7	0.7717	0.7680	1.4298	434.85	14.80	888.49	0.2234	0.6863
16	n-Hexadecane	$C_{16}H_{34}$	16	226.4	18.2	286.9	0.7730	0.7729	1.4325	449.85	14.00	944.33	0.2199	0.7174
17	n-Heptadecane	$C_{17}H_{36}$	17	240.5	22.0	302.2	0.7752	0.7765	1.4348	462.85	13.40	999.83	0.2189	0.7697
18	n-Octadecane	$C_{18}H_{38}$	18	254.5	28.2	316.7	0.7841	0.7805	1.4369	473.85	12.70	1059.74	0.2167	0.8114
19	n-Nonadecane	$C_{19}H_{40}$	19	268.5	31.9	329.9	0.7880	0.7844	1.4388	484.85	12.10	1119.82	0.2150	0.8522
20	n-Eicosane	$C_{20}H_{42}$	20	282.6	36.4	343.8	0.7890	0.7871	1.4405	494.85	11.60	1169.50	0.2125	0.9069
21	n-Heneicosane	$C_{21}H_{44}$	21	296.6	40.2	356.5	0.7954	0.7906	1.4440	504.85	11.10	1229.41	0.2110	0.9420
22	n-Docosane	$C_{22}H_{46}$	22	310.6	44.0	368.6	0.7981	0.7929	1.4454	513.85	10.60	1289.49	0.2089	0.9722

TABLE 2.5

Constants of Equation 2.13 for Various Properties of Pure Hydrocarbons

θ	C No. Range	θ_∞	a	b	c
Constants for Physical Properties of n-Alkanes					
T_M	C_5–C_{40}	397	6.5096	0.14187	0.470
T_b	C_5–C_{40}	1070	6.98291	0.02013	2/3
SG	C_5–C_{19}	0.85	92.22793	89.82301	0.01
d_{20}	C_5–C_{40}	0.859	88.01379	85.7446	0.01
I	C_5–C_{40}	0.2833	87.6593	86.62167	0.01
$T_{br}=T_b/T_c$	C_5–C_{20}	1.15	−0.41966	0.02436	0.58
$-P_c$	C_5–C_{20}	0	4.65757	0.13423	0.5
d_c	C_5–C_{20}	0.26	−3.50532	1.5×10^{-6}	2.38
$-\omega$	C_5–C_{20}	0.3	−3.06826	−1.04987	0.2
σ	C_5–C_{20}	33.2	5.29577	0.61653	0.32
Constants for Physical Properties of n-Alkylcyclopentanes					
T_M	C_7–C_{41}	370	6.52504	0.04945	2/3
T_b	C_6–C_{41}	1028	6.95649	0.02239	2/3
SG	C_7–C_{25}	0.853	97.72532	95.73589	0.01
d_{20}	C_5–C_{41}	0.857	85.1824	83.65758	0.01
I	C_5–C_{41}	0.283	87.55238	86.97556	0.01
$T_{br}=T_b/T_c$	C_5–C_{18}	1.2	0.06765	0.13763	0.35
$-P_c$	C_6–C_{18}	0	7.25857	1.13139	0.26
$-d_c$	C_6–C_{20}	−0.255	−3.18846	0.1658	0.5
$-\omega$	C_6–C_{20}	0.3	−8.25682	−5.33934	0.08
σ	C_6–C_{25}	30.6	14.17 95	7.02549	0.12
Constants for Physical Properties of n-Alkylcyclohexane					
T_M	C_7–C_{20}	360	6.55942	0.04681	0.7
T_b	C_6–C_{20}	1100	7.00275	0.01977	2/3
SG	C_6–C_{20}	0.845	−1.51518	0.05182	0.7
d_{20}	C_6–C_{21}	0.84	−1.58489	0.05096	0.7
I	C_6–C_{20}	0.277	−2.45512	0.05636	0.7
$T_{br}=T_b/T_c$	C_6–C_{20}	1.032	−0.11095	0.1363	0.4
$-P_c$	C_6–C_{20}	0	12.3107	5.53366	0.1
$-d_c$	C_6–C_{20}	−0.15	−1.86106	0.00662	0.8
$-\omega$	C_7–C_{20}	0.6	−5.00861	−3.04868	0.1
σ	C_6–C_{20}	31	2.54826	0.00759	1.0
Constants for Physical Properties of n-Alkylbenzenes					
T_M	C_9–C_{42}	375	6.53599	0.04912	2/3
T_b	C_6–C_{42}	1015	6.91062	0.02247	2/3
$-SG$	C_6–C_{20}	−0.8562	224.7257	218.518	0.01
$-d_{20}$	C_6–C_{42}	−0.854	238.791	232.315	0.01
$-I$	C_6–C_{42}	−0.2829	137.0918	135.433	0.01
$T_{br}=T_b/T_c$	C_6–C_{20}	1.03	−0.29875	0.06814	0.5
$-P_c$	C_6–C_{20}	0	9.77968	3.07555	0.15
$-d_c$	C_6–C_{20}	−0.22	−1.43083	0.12744	0.5
$-\omega$	C_6–C_{20}	0	−14.97	−9.48345	0.08
σ	C_6–C_{20}	30.4	1.98292	−0.0142	1.0

The C No. Range heading and "Constants in Equation 2.13" span over columns.

* Units: T_M, T_b, and T_c are in K. d and d_c are in g/cm³. P_c is in bar. σ is in dyn/cm.

Characteristics and Properties of Petroleum

refractive index, carbon-to-hydrogen weight ratio, and viscosity. These properties can be estimated from readily measurable properties in the laboratory such as boiling and specific gravity through the following methods as recommended in ASTM Manual 50 (Riazi, 2007; Riazi and Daubert, 1987).

$$M = 42.965[\exp(2.097\times10^{-4}T_b - 7.78712SG + 2.08476\times10^{-3}T_bSG)]T_b^{1.26007}SG^{4.98308}$$

$$(2.14)$$

$$T_c = 9.5233[\exp(-9.314\times10^{-4}T_b - 0.544442SG + 6.4791\times10^{-4}T_bSG)]T_b^{0.81067}SG^{0.53691}$$

$$(2.15)$$

$$P_c = 3.1958\times10^{5}[\exp(-8.505\times10^{-3}T_b - 4.8014SG + 5.749\times10^{-3}T_bSG)]T_b^{-0.4844}SG^{4.0846}$$

$$(2.16)$$

where T_b is the average or mid-boiling point in K and SG is the specific gravity at 60°F. P_c is in bar and T_c in K and V_c is in cm³/mol. Simpler version of these equations are given below:

$$M = 1.6607\times10^{-4}T_b^{2.1962}SG^{-1.0164} \tag{2.17}$$

$$T_c = 19.06232T_b^{0.58848}SG^{0.3596} \tag{2.18}$$

$$P_c = 5.53027\times10^{7}T_b^{-2.3125}SG^{2.3201} \tag{2.19}$$

$$V_c = 1.7842\times10^{-4}T_b^{2.3829}SG^{-1.683} \tag{2.20}$$

When T_b is not available, it can be estimated from M and SG using the following relation:

$$T_b = 3.76587\,[\exp(3.7741\times10^{-3}M + 2.98404SG - 4.25288\times10^{-3}MSG)]\,M^{0.40167}SG^{-1.58262}$$

$$(2.21)$$

The acentric factor (ω) can be estimated from the Lee–Kesler method (Lee and Kesler, 1975):

$$\omega = \frac{-\ln\dfrac{P_c}{1.01325} - 5.92714 + 6.09648/T_{br} + 1.28862\ln T_{br} - 0.169347T_{br}^{6}}{15.2518 - 15.6875/T_{br} - 13.4721\ln T_{br} + 0.43577T_{br}^{6}} \tag{2.22}$$

where P_c is in bar and T_{br} is the reduced boiling point which is defined as: $T_{br}=T_b/T_c$.
    For fractions with $M > 200$, if T_b is not available, M can be estimated from viscosity data as:

$$M = 223.56 \left[v_{38(100)}^{(-1.2435+1.1228SG)} \, v_{99(210)}^{(3.4758-3.038SG)} \right] SG^{-0.6665} \tag{2.23}$$

The three input parameters are kinematic viscosities (in cSt) at 38 and 98.9°C (100 and 210°F) shown by $v_{38(100)}$ and $v_{99(210)}$, respectively, and the specific gravity at 15.5°C, SG. If SG is not available it can be estimated from viscosity as:

$$SG = 0.7717 \left[v_{38(100)}^{0.1157} \right] \times \left[v_{99(210)}^{-0.1616} \right] \tag{2.24}$$

Density and specific gravity may be estimated from each other depending on the availability of data:

$$d_{20} = SG - 4.5 \times 10^{-3} (2.34 - 1.9SG) \tag{2.25}$$

$$SG = 0.9915 d_{20} + 0.01044$$
$$SG = 0.9823 d_{25} + 0.02184 \tag{2.26}$$

The refractive index of liquid hydrocarbons at 20°C (n) can be calculated from Equation 2.11 as follows.

$$n = \left(\frac{1+2I}{1-I} \right)^{1/2} \tag{2.27}$$

Parameter I should be calculated from:

$$I = 0.3773 T_b^{-0.02269} SG^{0.9182} \tag{2.28}$$

where T_b is in K. The CH ratio which is needed to estimate the composition of a petroleum product is estimated from the boiling point and specific gravity as:

$$CH = 3.4707 \left[\exp\left(1.485 \times 10^{-2} T_b + 16.94SG - 1.2492 \times 10^{-2} T_b SG \right) \right]$$
$$T_b^{-2.725} SG^{-6.798} \tag{2.29}$$

The following equations for predicting the freezing point of n-alkanes (P), n-alkyl-cyclopentanes (N), and n-alkylbenzenes (A) from molecular weight can be used in the carbon range of C_5–C_{40}.

$$T_{MP} = 397 - \exp(6.5096 - 0.14187 M^{0.47}) \tag{2.30}$$

$$T_{MN} = 370 - \exp(6.52504 - 0.04945 M^{2/3}) \tag{2.31}$$

$$T_{MA} = 395 - \exp(6.53599 - 0.04912 M^{2/3}) \tag{2.32}$$

Characteristics and Properties of Petroleum

where T_M is the melting point in K. Generally viscosity is a readily measurable property and is considered as an available property in terms of kinematic viscosity at 100°F (37.78 ~ 38°C). However, when this property is not available it can be estimated from the following relations:

$$\log v_{39(100)} = 4.39371 - 1.94733 K_W + 0.12769 K_W^2$$

$$+ 3.2629 \times 10^{-4} API^2 - 1.18246 \times 10^{-2} K_W API$$

$$+ \frac{0.171617 K_W^2 + 10.9943(API) + 9.50663 \times 10^{-2}(API)^2 - 0.860218 K_W(API)}{(API) + 50.3642 - 4.78231 K_W}$$

(2.33A)

$$\log v_{99(210)} = -0.463634 - 0.166532(API) + 5.13447 \times 10^{-4}(API)^2 - 8.48995 \times 10^{-3} K_W API$$

$$+ \frac{8.0325 \times 10^{-2} K_W + 1.24899(API) + 0.19768(API)^2}{(API) + 26.786 - 2.6296 K_W}$$

(2.33B)

where $v_{38(100)}$ and $v_{99(210)}$ are the kinematic viscosities in cSt (mm²/s) at 100 and 210°F, respectively. K_W and API are defined by Equations 2.8 and 2.2.

For a crude oil and reservoir fluid which can be assumed as a mixture of single carbon number (SCN) groups, the properties of each SCN can be estimated from Equation 2.13 using the constants given in Table 2.6 (Riazi and Al-Sahhaf, 1996).

TABLE 2.6
Coefficients of Equation 2.13 for Physical Properties of SCN Groups ($\geq C_{10}$) in Reservoir Fluids and Crude Oils

		Constants in Equation 2.13		
θ	$\theta\infty$	a	b	c
T_b	1080	6.97996	0.01964	2/3
SG	1.07	3.56073	2.93886	0.1
d_{20}	1.05	3.80258	3.12287	0.1
I	0.34	2.30884	2.96508	0.1
$T_{br} = T_b/T_c$	1.2	−0.34742	0.02327	0.55
$-P_c$	0	6.34492	0.7239	0.3
$-d_c$	−0.22	−3.2201	0.0009	1.0
$-\omega$	0.3	−6.252	−3.64457	0.1
σ	30.3	17.45018	9.70188	0.1
δ	8.6	2.29195	0.54907	0.3

Units: T_b, T_c in K; P_c in bar; d_{20} and d_c in g/cm³, σ in dyne/cm; δ in [cal/cm³]$^{1/2}$.

Further information on the properties of crude oils and reservoir fluids is given in Chapter 4 of ASTM MNL50 (Riazi, 2007).

Crude oils and reservoir fluids which have a wide boiling range cannot be represented just by a few components, and they are best characterized through the use of a distribution model such as the following probability density function (PDF) (Riazi, 1997):

$$F(P^*) = \frac{B^2}{A} P^{*B-1} \exp\left(-\frac{B}{A} P^{*B} \right)$$ (2.34)

$$P^* = \frac{P - P_o}{P_o}$$

where A and B are constants specific for each oil and a property "P" such as absolute boiling point (T_b), molecular weight (M), or specific gravity (SG). Through using this PDF and the Gaussian quadrature method, a crude oil mixture can be represented by a number of pseudocomponents that best describe the mixture as outlined in ASTM MNL50 (Riazi, 2007).

2.4 ESTIMATION OF THE COMPOSITION OF PETROLEUM MIXTURES

Petroleum fractions and products may be modeled as a mixture of three pseudo-components from parrafins (P), naphthenes (N), and aromatic (A) groups having the same molecular weight as that of the fraction. Properties of this mixture then can be calculated from the following mixing rule:

$$\theta = x_P \theta_P + x_N \theta_N + x_A \theta_A$$ (2.35)

where θ is a physical property for the mixture and θ_P, θ_N, and θ_A are the values of θ for the model components from the three groups with the composition of x_P, x_N, and x_A. Methods of prediction of the PNA composition of a petroleum mixture are given by Riazi and Daubert (1986). In these methods, the PNA composition is calculated through characterization parameters such as R_i, m, SG, and CH where R_i is defined by Equation 2.9 and parameter m is defined in terms of M and n.

$$m = M(n - 1.475)$$ (2.36)

where M can be calculated by Equation 2.17 and parameters n, d, and CH can be calculated through Equations 25–29. Depending on the value of M, the composition (x_P, x_N, x_A) can be estimated from the following relations.

For fractions with $M \leq 200$:

$$x_P = 2.57 - 2.877SG + 0.02876CH$$ (2.37)

$$x_N = 0.52641 - 0.7494x_P - 0.021811m$$ (2.38)

or

Characteristics and Properties of Petroleum

$$x_P = 3.7387 - 4.0829SG + 0.014772m \qquad (2.39)$$

$$x_N = -1.5027 + 2.10152SG - 0.02388m \qquad (2.40)$$

For fractions with $M > 200$:

$$x_P = 1.9842 - 0.27722R_i - 0.15643CH \qquad (2.41)$$

$$x_N = 0.5977 - 0.761745R_i + 0.068048CH \qquad (2.42)$$

or

$$x_P = 1.9382 + 0.074855m - 0.19966CH \qquad (2.43)$$

$$x_N = -0.4226 - 0.00777m + 0.107625CH \qquad (2.44)$$

If the calculated value of x_P or x_N is negative it should be assumed as zero. Regardless of which method is used, x_A must be calculated from the following equation.

$$x_A = 1 - (x_P + x_N) \qquad (2.45)$$

In addition to PNA composition, aromatics are divided into two groups of mono-aromatics (x_{MA}) and polyaromatics (x_{PA}) which may be estimated through the following relations for fractions with $M < 250$.

$$x_{MA} = -62.8245 + 59.90816R_i - 0.0248335m \qquad (2.46)$$

$$x_{PA} = 11.88175 - 11.2213R_i + 0.023745m \qquad (2.47)$$

$$x_A = x_{MA} + x_{PA} \qquad (2.48)$$

The average error for Equations 2.37 through 2.48 is about 0.05 or 5%.

2.5 ESTIMATION OF DENSITY AND POUR POINT

As discussed in Section 2.4, generally the API gravity or specific gravity is available along with another property such as boiling point or kinematic viscosity. The density of a fluid (gases and liquids) is a function of temperature, pressure, and composition. However, the effect of pressure on the density of gases and the effect of composition on the density of liquids are more significant. Density may be expressed in the form of absolute density (ρ, g/cm³), molar volume (V, cm³/mol), or compressibility factor ($Z = PV/RT = V/V^{ig}$, dimensionless). Z can be estimated from equations of states or generalized correlations at a given T and P for any fluid, and density is calculated from:

$$\rho = \frac{MP}{ZRT} \qquad (2.49)$$

where M is the molecular weight, R is the gas constant, and T is the absolute temperature. If M is in g/mol, P in bars, T in K, and $R = 83.14$ bar-cm^3/mol-K, then ρ is calculated in g/cm^3 or kg/L.

The density of gases generally can be estimated more accurately than the density of liquids, and one simple method is the truncated virial equation with three terms:

$$Z = 1 + \frac{B}{V} + \frac{C}{V^2} \tag{2.50}$$

This equation can be converted into another equivalent version in terms of P as:

$$Z = 1 + (B/RT)P + (C - B^2)(P/RT)^2 \tag{2.51}$$

which Z can be directly calculated at a given T and P. Parameters B and C are the second and third virial coefficients which are functions of temperature and can be calculated from the following generalized correlations (Equations 2.52 and 2.53) in terms of reduced temperature $T_r = T/T_c$:

$$\frac{BP_c}{RT_c} = B^{(0)} + \omega B^{(1)}$$

$$B^{(0)} = 0.1445 - \frac{0.330}{T_r} - \frac{0.1385}{T_r^2} - \frac{0.0121}{T_r^3} - \frac{0.000607}{T_r^8} \tag{2.52}$$

$$B^{(1)} = 0.0637 + \frac{0.331}{T_r^2} - \frac{0.423}{T_r^3} - \frac{0.008}{T_r^8}$$

$$\frac{CP_c^2}{(RT_c)^2} = C^{(0)} + \omega C^{(1)}$$

$$C^{(0)} = 0.01407 + \frac{0.02432}{T_r^{2.8}} - \frac{0.00313}{T_r^{10.5}} \tag{2.53}$$

$$C^{(1)} = -0.02676 + \frac{0.0177}{T_r^{2.8}} + \frac{0.040}{T_r^3} - \frac{0.003}{T_r^6} - \frac{0.00228}{T_r^{10.5}}$$

For natural gases the Hall–Yarborough (1971) method may be used as given below:

$$Z = 0.06125 P_r T_r^{-1} y^{-1} \exp\left[-1.2\left(1 - T_r^{-1}\right)^2\right] \tag{2.54}$$

Characteristics and Properties of Petroleum

where T_r and P_r are reduced temperature and pressure and y should be obtained from the solution of the following equation:

$$F(y) = -0.06125 P_r T_r^{-1} \exp\left[-1.2\left(1-T_r^{-1}\right)^2\right] + \frac{y+y^2+y^3-y^4}{\left(1-y\right)^3}$$

$$-\left(14.76 T_r^{-1} - 9.76 T_r^{-2} + 4.58 T_r^{-3}\right) y^2 \tag{2.55}$$

$$+\left(90.7 T_r^{-1} - 242.2 T_r^{-2} + 42.4 T_r^{-3}\right) y^{\left(2.18+2.82 T_r^{-1}\right)} = 0$$

The above equation can be solved by the Newton–Raphson method as shown by Riazi (2007) in ASTM Manual 50. For gas mixtures, T_c and P_c can be replaced with pseudo-critical properties T_{pc} and P_{pc} which may be calculated from the simple Kay's mixing rule:

$$T_{pc} = \sum_{i=1}^{N} x_i T_{ci}$$

$$\tag{2.56}$$

$$P_{pc} = \sum_{i=1}^{N} x_i P_{ci}$$

For gases at atmospheric pressure, Equation 2.49 adequately calculates gas density with $Z=1$, and M can be obtained from SG_g, the gas specific gravity, through Equation 2.3. If gas composition is known, the molecular weight can be calculated from the mixing rule, $M = \sum x_i M_i$.

For pure liquids, density can be calculated directly from specific gravity (SG) through Equations 2.1, 2.24, and 2.26 at a specified temperature. Density at other temperatures and at 1 atmosphere can be calculated from the density at one reference temperature T_o (Riazi, 2007):

$$d_T = d_{To} - 10^{-3} \times (2.34 - 1.9 d_T) \times (T - T_o) \tag{2.57}$$

where both T and T_o are in K or in °C and d_T and d_{To} are in g/cm³. If SG of oil is the only available data, then Equation. 2.57 becomes:

$$\rho_T{}^o = 0.999 SG - 10^{-3} \times \left(2.34 - 1.898 SG\right) \times \left(T - 288.7\right) \tag{2.58}$$

where SG is the specific gravity at 15.5°C (60°F/60°F) and T is absolute temperature in K. ρ_T^o is liquid density at temperature T and atmospheric pressure in g/cm³.

The density of liquid water at 1 atm (as a reference fluid), for temperatures in the range of 0 to 60°C, is given by AIChE-DIPPR (1996) as:

$$d_T = A \times B^{-\left[1+\left(1-T/C\right)^D\right]} \tag{2.59}$$

where T is in K and d_T is the density of water at temperature T in g/cm³. The coefficients are: $A = 9.83455 \times 10^{-2}$, $B = 0.30542$, $C = 647.13$, $D = 0.081$. The accuracy of this equation is within 0.1%.

36 Oil Spill Occurrence, Simulation, and Behavior

For liquid mixtures, if the composition is known, density can be calculated using the following mixing rule

$$\frac{1}{\rho_{mix}} = \sum_i \frac{x_{wi}}{\rho_i} \qquad (2.60)$$

where ρ_{mix} is the mixture liquid density (i.e., g/cm³) and x_{wi} is the weight fraction of component i in the liquid mixture. ρ_i should be known from a database, experiment, or may be calculated from the Rackett equation. For saturated liquids, the molar volume (V^{sat}) can be calculated from the Rackett equation with parameters given by Spencer and Danner (1972):

$$V^{sat} = \left(\frac{RT_c}{P_c}\right) Z_{RA}^{\ n}$$

$$n = 1.0 + \left(1.0 - T_r\right)^{2/7} \qquad (2.61)$$

If the Rackett parameter (Z_{RA}) is not available, it may be calculated from density at one temperature, d_T:

$$Z_{RA} = \left(\frac{MP_c}{RT_c d_T}\right)^{1/n} \qquad (2.62)$$

where n should be calculated from Equation 2.61 at temperature T at which density is known. Other methods for the calculation of Z_{RA} are given in ASTM MNL50.

The effect of pressure on liquid density is not as significant as the effect of pressure on the density of gases, and for this reason for small changes in pressure this effect can be neglected. The density of liquids at high pressures can be estimated from the density at lower pressure and the same temperature using the API-recommended method (API-TDB, 1997):

$$\frac{\rho^\circ}{\rho} = 1.0 - \frac{P}{B_T} \qquad (2.63)$$

where ρ° is the liquid density at low pressures (atmospheric pressure) and ρ is the density at high pressure P (in bar). B_T is calculated from the following set of equations:

$$B_T = mX + B_l$$

$$m = 1492.1 + 0.0734P + 2.0983 \times 10^{-6} P^2$$

$$X = \left(B_{20} - 10^5\right)/23170 \qquad (2.64)$$

$$\log B_{20} = -1.098 \times 10^{-3} T + 5.2351 + 0.7133\rho^\circ$$

$$B_l = 1.0478 \times 10^3 + 4.704P - 3.744 \times 10^{-4} P^2 + 2.2331 \times 10^{-8} P^3$$

Characteristics and Properties of Petroleum

where B_T is in bar and ρ° is the liquid density at atmospheric pressure in g/cm³. In the above equation T is absolute temperature in degrees Kelvin and P is the pressure in bar. If ρ° is not available, it can be estimated from Equation 2.58 or through the Rackett method. There are some other methods to calculate the density of liquid petroleum fractions at high pressures which are discussed in ASTM MNL50 (Riazi, 2007).

In order to have a flow of oil in pipelines, the pour point of oil should be less than the ambient temperature. It can be measured by ASTM D-97 (ISO 3016 or IP 15) test methods and may be estimated from viscosity and specific gravity through the following relation:

$$T_P = 130.47\,[SG^{2.970566}]\times[M^{(0.61235-0.47357\,SG)}]\times[v_{38(100)}^{(0.310331-0.32834\,SG)}] \qquad (2.64)$$

where T_P is the pour point in Kelvin, M is the molecular weight, and $v_{38(100)}$ is the kinematic viscosity at 37.8°C (100°F) in cSt. When M and $v_{38(100)}$ are not available they may be estimated through the methods discussed earlier in this section. Heavier oils with higher wax contents have higher pour points.

2.6 ESTIMATION OF BOILING POINT, FLASH POINT, AND VAPOR PRESSURE

Boiling point (T_b), vapor pressure (P^{vap}), and flash point (T_F) are all related to oil volatility and are important in order to determine how fast a petroleum product can vaporize. The boiling point can be calculated from Equation 2.21 from values of M and SG. Experimentally it can be determined by ASTM D86 or by simulated distillation using ASTM D2887 test methods. If the temperature at 10 vol% vaporized by ASTM D86 is given by T_{10}, then the flash point can be estimated from the following relation:

$$T_F = 15.48 + 0.70704\,T_{10} \qquad (2.65)$$

where both T_{10} and T_F are in degrees Kelvin. More recently Alqaheem and Riazi (2017) have shown that for hydrocarbons and their mixtures the ratio of flash point to normal boiling point is almost constant and equal to 0.7 when both temperatures are in absolute degrees:

$$T_F = 0.7\,T_b \qquad (2.66)$$

This method should be used with caution and provides approximate values of the flash point.

There are many methods to estimate the vapor pressure of hydrocarbons and petroleum fractions as discussed in ASTM Manual 50 (Riazi, 2007). However, one of the simplest methods based on corresponding state theory was developed by Lee and Kesler (1975) which requires T_b, T_c, P_c, and acentric factor (ω) as given below.

$$\ln P_r^{vp} = 5.92714 - \frac{6.09648}{T_r} - 1.28862 \ \ln T_r + 0.169347 T_r^6$$

$$+ \omega \left(15.2518 - \frac{15.6875}{T_r} - 13.4721 \ \ln T_r + 0.43577 T_r^6 \right) \tag{2.67}$$

where $P_r^{vp} = P^{vp}/P_c$ and $T_{br} = T_b/T_c$.

Another simple method for the quick estimation of vapor pressure at temperatures near the boiling point when T_b is the only available data is given as:

$$\log_{10} P^{vap} = 3.2041 \left(1 - 0.9982 \times \frac{T_b - 41}{T - 41} \times \frac{1393 - T}{1393 - T_b} \right) \tag{2.68}$$

The units for T and T_b is K, and P is in bar.

2.7 ESTIMATION OF VISCOSITY, DIFFUSIVITY, AND SURFACE TENSION

Viscosity and surface tension are physical properties which are important in determining how fast an oil spill spreads on seawater surface. Part of the oil dissolves in water or vaporizes into air, and it is the diffusivity or diffusion coefficient of oil molecules in water or air that determines how fast oil particles dissolve in water or vaporize into air. All these properties depend on oil composition, temperature, and pressure, and usually they can be estimated through the density of oil.

A simplified method to get an estimate of the viscosity of undefined hydrocarbon gas mixtures at atmospheric pressure is given in terms of M and T by API-TDB (1997) as:

$$\mu_o^g = -0.0092696 + \sqrt{T} \left(0.001383 - 5.9712 \times 10^{-5} \sqrt{M} \right) + 1.1249 \times 10^{-5} M \tag{2.69}$$

where T is in K and μ_o^g is the viscosity of gas at low pressure in cp. An empirical relation for the calculation of the viscosity of natural gases at any T and P was proposed by Lee et al. (1966):

$$\mu^g = 10^{-4} A \left[\exp \left(B \times \rho^C \right) \right]$$

$$A = \left[(12.6 + 0.021M) T^{1.5} \right] / \left(116 + 10.6M + T \right)$$

$$B = 3.45 + 0.01M + \frac{548}{T} \tag{2.70}$$

$$C = 2.4 - 0.2B$$

where μ^g is the viscosity in cp, M is molecular weight, T is gas temperature in K, and ρ is the gas density in g/cm^3 at the same conditions of the gas. This equation may be used up to a pressure of 550 bar.

Characteristics and Properties of Petroleum

For the estimation of viscosity of dense hydrocarbons and reservoir fluids, the method proposed by Jossi et al. (1962) may be used.

$$\left[\left(\mu-\mu_\circ\right)\xi+10^{-4}\right]^{\frac{1}{4}}=0.1023+0.023364\rho_r+0.058533\rho_r^2-0.040758\rho_r^3+0.0093324\rho_r^4 \tag{2.71}$$

where μ is in cp, and T_r is the reduced temperature and ξ is a parameter with dimension of reciprocal viscosity and is defined as:

$$\xi = T_c^{\frac{1}{6}} M^{-\frac{1}{2}} \left(0.987 P_c\right)^{-\frac{2}{3}} \tag{2.72}$$

In the above relation, $(\mu - \mu_a)\,\xi$ is dimensionless and ρ_r is the reduced density ($\rho_r = \rho/\rho_c = V_c/V$). μ_o is viscosity at low pressure and could be calculated from the following method developed earlier by Stiel and Thodos and recommended in the API-TDB (1997) to calculate the viscosity of gases at low pressures:

$$\mu\xi = 3.4\times10^{-4}T_r^{0.94} \qquad\qquad\qquad \text{for } T_r \le 1.5$$
$$\mu\xi = 1.778\times10^{-4}\left(4.58T_r-1.67\right)^{0.625} \qquad \text{for } T_r > 1.5 \tag{2.73}$$

where units of μ and ξ are the same as in Equations 2.71–2.73.

For crude oils at atmospheric pressure, kinematic viscosity ν at any T may be estimated from the following method suggested by API-TDB (1997):

$$\log\left(\nu_T\right) = A\left(\frac{311}{T}\right)^B - 0.8696$$

$$A = \log\left(\nu_{39(100)}\right)+0.8696 \tag{2.74}$$

$$B = 0.28008\times\log\left(\nu_{39(100)}\right)+1.8616$$

where T is temperature in K and $\nu_{39(100)}$ is the kinematic viscosity at 100°F (39°C or 311 K) in cSt which is usually known from experiment or could be calculated from Equation 2.34.

Another method is Glaso's correlation for the viscosity of crude oils at atmospheric pressure:

$$\mu_{od} = \left(3.141\times10^{10}\right)\times\left[\left(1.8T-460\right)^{-3.444}\right]\times\left[\log_{10}(API)\right]^n \tag{2.75}$$

$$n = 10.313\left[\log_{10}(1.8T-460)\right]-36.447$$

where μ_{od} is the viscosity of crude oil (gas free at 1 atm) in cp, T is temperature in K, and API is the oil gravity. The Kouzel correlation may be used to estimate the viscosity of liquid oil at high pressures:

$$\log \frac{\mu_P}{\mu_a} = \frac{P - 1.0133}{10000} \left(-1.48 + 5.86 \mu_a^{0.181} \right) \tag{2.76}$$

where P is pressure in bar and μ_a is viscosity at low pressure (1 atm) in cp.

For the estimation of diffusion coefficients in gases at low pressures, the well-known Chapman–Enskog equation is suitable with reasonable accuracy:

$$\left(\rho D_{AB} \right)^o = \frac{2.2648 \times 10^{-5} T^{0.5} \left(\dfrac{1}{M_A} + \dfrac{1}{M_B} \right)^{0.5}}{\sigma_{AB}^2 \Omega_{AB}}$$

$$\sigma_{AB} = \frac{\sigma_A + \sigma_B}{2}$$

$$\sigma_i = 0.1866 V_{ci}^{1/3} Z_{ci}^{-6/5}$$

$$\Omega_{AB} = \frac{1.06036}{\left(T_{AB}^* \right)^{0.1561}} + 0.193 \exp\left(-0.47635 T_{AB}^* \right) + 1.76474 \exp\left(-3.89411 T_{AB}^* \right) + \tag{2.77}$$

$$1.03587 \exp\left(-1.52996 T_{AB}^* \right)$$

$$T_{AB}^* = T / \varepsilon_{AB}$$

$$\varepsilon_{AB} = \left(\varepsilon_A \varepsilon_B \right)^{1/2}$$

$$\varepsilon_i = 65.3 T_{ci} Z_{ci}^{18/5}$$

where $(\rho D_{AB})^o$ represents the product of the density-diffusivity of an ideal gas at low pressure in the unit of mol/cm·s and D_{AB}^o would be in cm²/s. ε and σ are the energy and size parameters, T and T_c are in K, and V_c is in cm³/mol. According to this method, the diffusion coefficient of component A in B at low pressure can be calculated through $D_{AB}^o = (\rho D_{AB})^o / \rho$ where ρ low pressure density may be calculated from ($\rho = 83.14 T/P$) in which T is in K and P is in bar.

There are some other empirical correlations for the estimation of the diffusion coefficient of light gases in reservoir fluids. For example, Renner (1988) proposed the following empirical correlation for the calculation of gases in oils under pressure:

$$D_{A-oil} = 10^{-9} \mu_{oil}^{-0.4562} M_A^{-0.6898} \rho_{MA}^{1.706} P^{-1.831} T^{4.524} \tag{2.78}$$

Characteristics and Properties of Petroleum

where $D_{A\text{-}oil}$ is the diffusivity of light gas A (C_1, C_2, C_3, CO_2) in an oil (reservoir fluid) in cm^2/s. μ_{oil} is the oil viscosity at T and P in cp, M_A is molecular weight of gas A, ρ_{MA} is molar density of gas at T and P in mol/cm^3, P is pressure in bar and T is absolute temperature in K. A more generalized method for the estimation of diffusion coefficients in reservoir fluids was proposed by Riazi and Whitson (1993) as below:

$$\frac{(\rho D_{AB})}{(\rho D_{AB})^\circ} = a \left(\frac{\mu}{\mu^\circ} \right)^{b + cP_r}$$

$$a = 1.07 \qquad\qquad b = -0.27 - 0.38\omega \qquad\qquad (2.79)$$

$$c = -0.05 + 0.1\omega \qquad\qquad P_r = P / P_c$$

$$T_c = x_A T_{cA} + x_B T_{cB} \qquad\qquad P_c = x_A P_{cA} + x_B P_{cB} \qquad\qquad \omega = x_A \omega_A + x_B \omega_B$$

where $(\rho D_{AB})^\circ$ must be determined from Equation 2.77. μ is the viscosity of the mixture at T and P of the system while μ° is viscosity at T and 1 atm pressure which can be calculated from Equation 2.73. In Equation 2.79, component A is the diffusing component and B is a mixture of all other components, the properties of which can be calculated from the mixing rule excluding component A as discussed in ASTM MNL50 (Riazi, 2007).

Finally in this part methods of calculating surface tension for pure compounds and interfacial tension between oil and water are presented. Surface tension (σ) is a molecular property that depends on the density of the liquid (ρ^L) and its vapor (ρ^V) at a given temperature and pressure. There are a number of predictive methods which correlate σ to temperature and critical properties. For example the Block and Bird correlation is given for pure hydrocarbons:

$$\sigma = P_c^{2/3} T_c^{1/3} Q \left(1 - T_r \right)^{11/9}$$

$$Q = 0.1196 \left[1 + \frac{T_{br} \ln(P_c / 1.01325)}{1 - T_{br}} \right] - 0.279 \qquad\qquad (2.80)$$

where σ is in dyn/cm, P_c in bar, T_c in K, and T_{br} is the reduced boiling point (T_b/T_c). This equation is not suitable for non-hydrocarbons. For undefined petroleum fractions the following method can be used for the calculation of surface tension:

$$\sigma = \frac{673.7 \left(1 - T_r \right)^{1.232}}{K_W} \qquad\qquad (2.81)$$

where T_r is the reduced temperature and K_W is the Watson characterization factor. Another method for the surface tension of hydrocarbons, petroleum fractions, and coals liquids is given below:

Oil Spill Occurrence, Simulation, and Behavior

$$\sigma^{1/4} = \frac{P_a}{M}\left(\rho^L - \rho^V\right)$$

$$\frac{P_a}{M} = 1.7237 T_b^{0.05873} SG^{-0.64927}$$

(2.82)

where T_b is the boiling point in K and SG is the specific gravity. The interfacial tension (IFT) between a hydrocarbon (or a petroleum fraction) and water can be calculated from the following method, proposed by Firoozabadi (1999) in the following form:

$$\sigma_{HW} = 111\left(\rho_W - \rho_H\right)^{1.024}\left(T / T_{cH}\right)^{-1.25}$$

(2.83)

where σ_{HW} is the hydrocarbon–water IFT in dyn/cm (mN/m), ρ_W and ρ_H are water and hydrocarbon densities in g/cm^3, T is temperature in K, and T_{cH} is the critical temperature of pure hydrocarbon (or petroleum fraction) in K. An error as high as 30% might be observed from this method.

2.8 SOLUBILITY OF HYDROCARBONS AND PETROLEUM FLUIDS IN WATER

As, during oil spill accident, oil and water come into contact with each other, the amount of solubility of oil or its components in water is important in the modeling of oil spill behavior in the sea. The amount of solubility of hydrocarbons in water is small, but from toxicological and environmental points of view the hydrocarbon solubility in water plays an important role. Solubility can be estimated from the general equilibrium relation between vapor and liquid phases (ASTM MNL50):

$$x_1 = \frac{\phi_1^V P}{\gamma_1 f_1^L}$$

(2.84)

where ϕ_1^V is the fugacity coefficient of pure gas (component 1) at T and P which at low pressure could be considered as unity. γ_1 is the activity coefficient of solute 1 in solvent 2 and f_1^L is the fugacity of pure component 1, which at low pressure could be replaced with vapor pressure (P^{vap}) at T. When oil and water are in contact, the principle of liquid–liquid equilibria (LLE) applies as below:

$$\hat{f}_i^{\text{water}} = \hat{f}_i^{\text{oil}}$$

(2.85)

where $\hat{f}_i^{\text{water}}$ is the fugacity of component i in water phase. Since the solubility of hydrocarbons in water is very low, we can assume that the water phase is

Characteristics and Properties of Petroleum

pure water. $\hat{f}_i^{oil}$ is the fugacity of i in the oil phase which can be calculated from:

$$\hat{f}_i^L = x_i \gamma_i f_i^L \tag{2.86}$$

where γ_i is the activity coefficient of component i in oil or in water phase and can be calculated from the solubility parameter.

$$\ln \gamma_i = \frac{V_i^L \left(\delta_i - \delta_{mix}\right)^2}{RT}$$

$$\delta_{mix} = \sum_j \Phi_j \delta_j \tag{2.87}$$

$$\Phi_j = \frac{x_j V_j^L}{\sum_k x_k V_k^L}$$

In which V_i is the liquid molar volume and δ is the solubility parameter at 25°C which may be calculated for n-alkanes (P), n-alkylcyclohexanes (N), and n-alkylbenzenes (A) from the following relations (Riazi, 2007):

$$\ln V_{25} = -0.51589 + 2.75092 M^{0.15} \qquad \text{for n-alkanes } (C_1 - C_{36})$$

$$V_{25} = 10.969 + 1.1784 M \qquad \text{for n-alkylcyclohexanes } (C_6 - C_{16}) \tag{2.88}$$

$$\ln V_{25} = -96.3437 + 96.54607 M^{0.01} \qquad \text{for n-alkylbenzenes } (C_6 - C_{24})$$

$$\delta = 16.22609 \left[1 + \exp\left(0.65263 - 0.02318 M\right)\right]^{-0.4007} \quad \text{for n-alkanes } (C_1 - C_{36})$$

$$\delta = 16.7538 + 7.2535 \times 10^{-5} M \qquad \text{for n-alkylcyclohexanes } (C_6 - C_{16})$$

$$\delta = 26.8557 - 0.18667 M + 1.36926 \times 10^{-3} M^2$$

$$- 4.3464 \times 10^{-6} M^3 + 4.89667 \times 10^{-9} M^4 \qquad \text{for n-alkylbenzenes } (C_6 - C_{24})$$

$$\tag{2.89}$$

where V_{25} is in cm³/mol and δ is in (J/cm³)$^{1/2}$.

For the solubility of hydrocarbons in water or water in oil due to low solubility, Henry's law can be used as below:

$$y_i P = k_i x_i \tag{2.90}$$

TABLE 2.7
Constants for Equation 2.91 for Estimation of Henry's Constant for Light Gases in Water

Gas	T Range, K	Pressure Range, bar	A_1	A_2	A_3	A_4	%AAD
Methane	274–444	1–31	569.29	0.107305	−19537	−92.17	3.6
Ethane	279–444	1–28	109.42	−0.023090	−8006.3	−11.467	7.5
Propane	278–428	1–28	1114.68	0.205942	−39162.2	−181.505	5.3
n-Butane	277–444	1–28	182.41	−0.018160	−11418.06	−22.455	6.2
i-Butane	278–378	1–10	1731.13	0.429534	−52318.06	−293.567	5.3

where k_i is Henry's constant and has the unit of pressure and for any given solute and solvent system is a function of temperature. Henry's constant for light hydrocarbon gases (C_1, C_2, C_3, C_4, and iC_4) in water may be estimated from the following correlation as suggested by the API-TDB (1997):

$$\ln k_{\text{gas-water}} = A_1 + A_2 T + \frac{A_3}{T} + A_4 \ln T \tag{2.91}$$

where T is in Kelvin and the coefficients A_1–A_4 and the ranges of T and P are given in Table 2.7.

For each gas the range of temperatures at which the method is applicable is given inside the parenthesis in degrees Kelvin. T is the absolute temperature in Kelvin and x is the mole fraction of dissolved gas in water at 1.013 bar. API-TDB (1997) recommends the following equation for the calculation of the solubility of water in some undefined petroleum fractions:

$$
\begin{aligned}
\text{Naphtha} \qquad & \log x_{\text{H2O}} = 2.94 + \frac{1841.3}{T} \\[1em]
\text{Kerosine} \qquad & \log x_{\text{H2O}} = 2.74 + \frac{2387.3}{T} \\[1em]
\text{Paraffinic Oil} \qquad & \log x_{\text{H2O}} = 2.69 + \frac{1708.3}{T} \\[1em]
\text{Gasoline} \qquad & \log x_{\text{H2O}} = 2.63 + \frac{1766.8}{T}
\end{aligned}
\tag{2.92}
$$

In the above equations, T is in degrees Kelvin and x_{H2O} is the mole fraction of water in the petroleum fraction.

Finally the solubility of light gases in water at 1 atmosphere can be estimated from the following correlations as given by Sandler (1999).

Characteristics and Properties of Petroleum

Methane (275-328) $\quad \ln x = -416.159289 + \dfrac{15{,}557.5631}{T} + 65.2552591 \ln T - 0.0616975729T$

Ethane (275-323) $\quad \ln x = -11{,}268.4007 + \dfrac{221{,}617.099}{T} + 2158.421791 \ln T - 7.18779402T$

$$+ 4.0501192 \times 10^{-3} T^2$$

Propane (273-347) $\quad \ln x = -316.46 + \dfrac{15921.2}{T} + 44.32431 \ln T$

n-Butane (276-349) $\quad \ln x = -290.238 + \dfrac{15{,}055.5}{T} + 40.1949 \ln T$

i-Butane (278-343) $\quad \ln x = 96.1066 - \dfrac{2472.33}{T} - 17.3663 \ln T$

H_2S (273-333) $\quad \ln x = -149.537 + \dfrac{8226.54}{T} + 20.2308 \ln T + 0.00129405T$

CO_2 (273-373) $\quad \ln x = -4957.824 + \dfrac{105{,}288.4}{T} + 933.17 \ln T - 2.854886T + 1.480857 \times 10^{-3} T^2$

N_2 (273-348) $\quad \ln x = -181.587 + \dfrac{8632.129}{T} + 24.79808 \ln T$

H_2 (274-339) $\quad \ln x = -180.054 + \dfrac{6993.54}{T} + 26.3121 \ln T - 0.0150432T$

$$(2.93)$$

the number in each parenthesis is the temperature range in degrees Kelvin. T is in Kelvin and x is the mole fraction of dissolved gas in water at 1.013 bar. Other methods for the calculation of the solubility of oil in water are discussed in ASTM MNL50 (Riazi, 2007).

2.9 SUMMARY

In this chapter, after the definition and classification of various oils, methods for physical and thermodynamic properties of pure hydrocarbons, crude oils, and petroleum products that might be needed in calculations related to predicting the fate of an oil spill were presented. The nature of crude oils and petroleum products, the composition, and characterization methods have been discussed. Methods for the estimation of the critical constants, molecular weight, boiling point, density, flash point, viscosity, diffusion coefficient, vapor pressure, surface tension, and solubility of hydrocarbons and petroleum mixtures in water were presented. In these methods, minimum available laboratory data can be used to estimate each property.

REFERENCES

AIChE DIPPR® Database. 1996. *Design Institute for Physical Property Data (DIPPR).* EPCON International, Houston.

Alqaheem, Sara S. and M.R. Riazi. 2017. Flash points of hydrocarbons and petroleum products, prediction and evaluations. *Energy & Fuels*, 31(4), pp. 3578–3584. http://pubs.acs.org/doi/abs/10.1021/acs.energyfuels.6b02669?src=recsys

API-TDB. 1997. *API technical data book - petroleum refining*, edited by T.E. Daubert and R.P. Danner, 6th Edition. American Petroleum Institute (API), Washington, DC.

Coplen, T.B. 2001. Atomic weights of the elements 1999. *Pure Applied Chemistry*, 73(4), pp. 667–683.

Firoozabadi, A. 1999. *Thermodynamics of hydrocarbon reservoirs.* McGraw Hill, New York.

Hall, K.R. and L. Yarborough. 1971. New simple correlation for predicting critical volume. *Chemical Engineering*, November, 78(25), pp. 76–77.

Jossi, J.A., L.I. Stiel and G. Thodos. 1962. The viscosity of pure substances in the dense gaseous and liquid phases. *AIChE Journal*, 8, pp. 59–63.

Lee, A., M.H. Gonzalez and B.E. Eakin. 1966. The viscosity of natural gases. *JPT*, 18(8), pp. 997–1000.

Lee, B.I. and M.G. Kesler. 1975. A generalized thermodynamic correlation based on three-parameter corresponding states. *American Institute of Chemical Engineers (AIChE) Journal*, 21, pp. 510–527.

Renner, T.A. 1988. Measurement and correlation of diffusion coefficients for oil and rich gas applications. *SPE Reservoir Engineering*, May, 3, pp. 517–523.

Riazi, M.R. 1997. A continuous model for C7+ characterization of petroleum fractions. *Industrial and Engineering Chemistry Research*, 36, pp. 4299–4307.

Riazi, M.R. 2007. *Characterization and properties of petroleum fractions.* MNL50, ASTM International, Conshohocken. ISBN: 978-0803133617. www.astm.org/DIGITAL_LIBRARY/MNL/SOURCE_PAGES/MNL50.htm, This book is also cited as ASTM MNL50 in this chapter.

Riazi, M.R. and T. Al-Sahhaf. 1995. Physical properties of n-Alkanes and n-Alkyl hydrocarbons: Application to petroleum mixtures. *Industrial and Engineering Chemistry Research*, 34, pp. 4145–4148.

Riazi, M.R. and T. Al-Sahhaf. 1996. Properties of heavy petroleum fractions and crude oils. *Fluid Phase Equilibria*, March 31, Elsevier, Holland, 117/1-2, pp. 217–224.

Riazi, M.R. and T.E. Daubert. 1986. Prediction of molecular type analysis of petroleum fractions and coal liquids. *Industrial & Engineering Chemistry, Process Design & Development*, 25(4), pp. 1009–1015.

Riazi, M.R. and T.E. Daubert. 1987. Characterization parameters for petroleum fractions. *Industrial & Engineering Chemistry Research*, 26, pp. 755–759.

Riazi, M.R. and C.H. Whitson. 1993. Estimating diffusion coefficients of dense fluids. *Industrial & Engineering Chemistry Research*, 32(12), pp. 3081–3088.

Sandler, S.I. 1999. *Chemical and engineering thermodynamics*, 3rd Edition. John Wiley and Sons, New York.

Speight, J.G. 1998. *The chemistry and technology of petroleum*, 3rd Edition. Marcel Dekker, Inc., New York.

Spencer, C.F. and R.P. Danner. 1972. Improved equation for prediction of saturated liquid density. *Journal of Chemical & Engineering Data*, 17, p. 236.

Watson, K.M., E.F. Nelson and G.B. Murphy. 1935. Characterization of petroleum fractions. *Industrial and Engineering Chemistry*, 27, pp. 1460–1464.

3 Sources and Causes of Oil Spills

ACRONYMS

API American Petroleum Institute
ASTM American Society for Testing and Materials
BP British Petroleum
CIS The Commonwealth of Independent States (from former Soviet Union)
EIA Energy Information Administration (part of US Department of Energy, Washington, DC)
EPA US Environmental Protection Agency
GOM Gulf of Mexico
GOR Gas-to-oil ratio
IEA International Energy Agency (based in Paris, France)
ITOPF International Tanker Owners Pollution Federation
LNG Liquefied natural gas
Mb/d Million barrels per day (1 b/d average oil is equivalent to 50 metric ton/ year)
MENA Middle East and North Africa
MNL Manual (published by ASTM International)
Mtoe Metric ton oil equivalent (7.33 barrel of oil)
NGL Natural gas liquid
NOAA National Oceanic and Atmospheric Administration
OECD Organization for Economic Cooperation and Development

3.1 INTRODUCTION

The UN Glossary of Statistical Terms defined oil spill as follows (OECD, 2001):

> An oil spill is oil, discharged accidentally or intentionally, that floats on the surface of water bodies as a discrete mass and is carried by the wind, currents and tides. Oil spills can be partially controlled by chemical dispersion, combustion, mechanical containment and adsorption. They have destructive effects on coastal ecosystems. Oil spills into rivers, seas and oceans are often caused by accidents in pipeline and storage facilities, refineries, tankers, drilling rigs and offshore production activities. These accidents are caused by human errors, equipment failure, natural disasters such as hurricanes and deliberate acts such as war, illegal dumper and terrorism. (OECD 2001)

The type and amount of the oil spilled on the water surface, along with weather conditions and the location of the oil spill, would determine how fast the oil spreads, evaporates, or sinks to the bottom of the sea. For example, heavy components in the oil have a greater tendency to sink than to vaporize. The solubility of oil in water also depends on its constituents. For example, aromatics are more soluble in water than paraffinic hydrocarbons, and heavier oils are less soluble in water than light oils. These factors also determine the methods of cleanup used to remove an oil spill.

The petroleum industry can be in general divided into upstream and downstream sectors. The upstream industry deals with exploration and the production of petroleum or natural gas from reservoir to the surface. This sector also includes exploratory wells which are used to test the availability of oil in a reservoir with both onshore and offshore technology. With advances in geological technology, the number of exploratory wells has been significantly reduced from a few decades ago. The downstream sector is the second major industry which is involved in the processing of petroleum and natural gas from crude oil to different petroleum products. Crude oil after production goes through field processes to separate oil, gas, and water from each other. The crude oil is then stored or transported to refinery sites or exported to other countries for processing. This sector often is referred to as the midstream industry. In other words, the midstream industry provides the link between the oil- and gas-producing areas and the population areas where the consumers are located. It is during offshore production and crude oil transportation by tankers that most oil spill accidents occur.

Gasoline as a motor fuel is the most valuable product from petroleum production. In the US alone more than 9 million barrels of gasoline per day are manufactured. A schematic of the petroleum industry in the US from the production to the distribution of gasoline is given in Figure 3.1 (EIA, 2013). In this figure the three sectors of

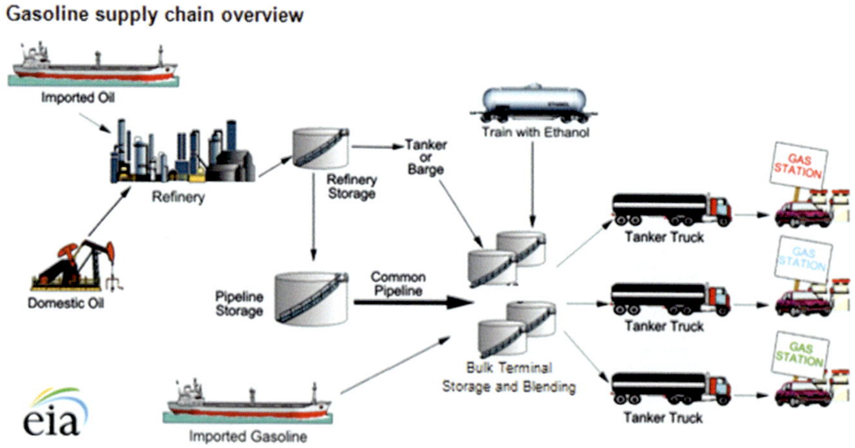

FIGURE 3.1   Schematic of petroleum industry in the US for gasoline production (EIA, 2013).

the industry, namely upstream, midstream, and downstream, are clearly shown from reservoir to gas stations.

In summary, in offshore production the upstream sector, which involves well, pipeline, platform, and ship, could be a potential source of oil spills. The midstream sector involves ship, truck, and storage, and the downstream sector includes oil and gas station, airplane, ship, car, train, and truck which are potential sources for the occurrence of oil spills. The sinking and breakup of ships are also other potential sources of oil spills in the marine environment.

3.2 STATISTICAL DATA ON OIL AND GAS PRODUCTION AND TRADE MOVEMENT

World energy consumption is on the rise, as shown in Figure 3.2 from 1993 to 2018 according to a BP statistical report (BP, 2019). Total energy consumption in the year 2018 was nearly 14,000 million tons oil equivalent (mtoe), of which oil consumption is about 4,750 mtoe (~95 Mb/d). The other two major energy sources after oil are natural gas and coal. In 2018 the world energy consumption grew by 2.9% from 2017, and the major countries for energy consumption growth were China, India, and the US. In 2019 oil consumption grew by an average of 2.2 Mb/d and natural gas consumption grew by 5.3%, the fastest rate since 1984.

According to the BP 2019 statistical report the oil production in the world within a decade from 2008 to 2018 increased by 14% from 83 to nearly 95 Mb/d (million barrels per day) as given in Table 3.1. However, EIA reports that world oil production combined with other liquids such as gas condensate in 2018 was 100.8 Mb/d. Almost 30% of world production is from offshore fields.

Figure 3.3 shows the amount of proven oil reservoirs in 1988, 2008, and 2018 in different parts of the world with an increase of 50% in 20 years. This increase is due

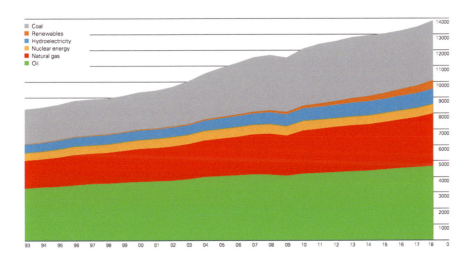

FIGURE 3.2 World energy consumption in million tons oil equivalent (mtoe) (BP, 2019).

TABLE 3.1
World Oil Production from 2008 to 2018 in Mb/d

Year	2008	2009	2010	2011	2012	2013	2014	2015	2016	2017	2018
Production	83.1	81.4	83.3	84	86.2	86.6	88.7	91.5	91.8	92.5	94.7

Source: BP.

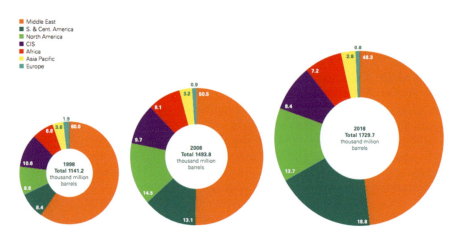

FIGURE 3.3 Percentage distribution of proved oil reserves in 1998, 2008, and 2018 (BP, 2019).

to the use of new technology in exploration and production as well as finding new recoverable oil reserves. The Middle East, South America, and North America have the largest world oil reserves. With respect to current production rates, the ratio of reserves to the rate indicates the lifetime of reserves in years as shown in Figure 3.4. Central and South America is the region with the highest R/P ratio, followed by the Middle East, while European oil production ends within a decade. The highest P/R ratio belongs to Central and South American countries with average lifetime of about 135 years. This is mainly due to the low production rate in Venezuela because of international sanctions although the country has vast oil reserves. With the current production rate, the Middle East region could run out of oil in about 75 years. Globally, with the 2018 production rate, it would take another 50 years of production for the world to run out of oil.

The daily rate of oil production and consumption in different parts of the world for the period of 1993 to 2018 is shown in Figure 3.5. North America's increase in production is attributed to the production of unconventional oil, and the largest increase in consumption is in the Asia Pacific region and in particular China.

Oil trade movements are mainly between producing and consuming countries. The total imports of crude oil in the world in 2008 were 56 million barrels per day, while this number in 2018 reached more than 71 million barrels daily (45.4 Mb/d crude oil

Sources and Causes of Oil Spills 51

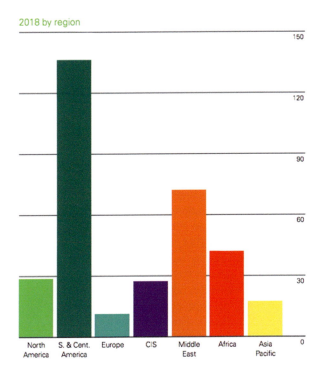

FIGURE 3.4 Reserves-to-production (R/P) ratios by region in 2018 in years (BP, 2019).

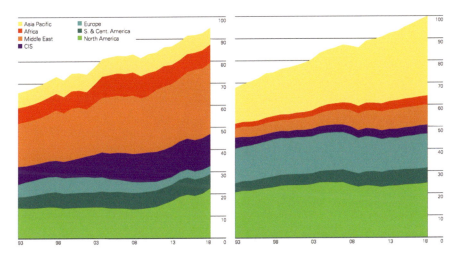

FIGURE 3.5 Oil production (left) and consumption (right) by region in million barrels daily (BP, 2019).

and 25.9 Mb/d products). The same amount of oil was exported from producing countries such as North America, Central and South America, Russia and CIS, the Middle East, and North Africa. The major oil importers are the US, Europe, China, India, and Japan. Major trade movements in 2018 are also shown in Figure 3.6.

Similarly for natural gas, the distribution of proven reserves is shown in Figure 3.7 with an increase of more than 50% from 1998 to 2018. The largest reserves are in the Middle East and CIS regions. The lifetime of natural gas reserves in different countries is shown in Figure 3.8.

As shown in Figure 3.8, the current global R/P ratio shows that gas reserves in 2018 accounted for 50.9 years of current production. The Middle East (109.9 years) and CIS (75.6 years) are the regions with the highest R/P ratios. Total world natural gas production in 2018 was 3,867.9 billion cubic meters (3,325.8 million tons oil equivalent or 66 Mb/d oil equivalent). The highest gas-producing country in the world was the US with 831.8 billion cubic meters. Natural gas production and consumption by region from 1993 to 2018 are shown in Figure 3.9 (BP, 2019).

Natural gas is transferred as liquefied natural gas (LNG), and in 2018 total LNG imports in the world were 431 billion cubic meters. Imports of natural gas by region are shown in Figure 3.10. The LNG trade movement between countries is shown in Figure 3.11.

Another type of liquid fuel whose production is growing rapidly is biofuels (mainly bio-ethanol and biodiesel). The total world production of biofuels in 2008 was 49.4 million tons oil equivalent (mtoe), which increased to 95.37 mtoe in 2018; production almost doubled in this 10-year period. From 2017 to 2018 the increase in production was almost 10% with Brazil and Indonesia accounting for two-thirds of global growth. North America, accounting for 56% of biodiesel production, is the world's top producer followed by Europe representing 37%. Worldwide production of biofuels is shown in Figure 3.12.

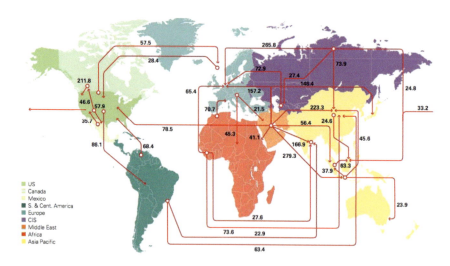

FIGURE 3.6 Major oil trade movements in 2018. Trade flows worldwide in million tons (BP, 2019).

Sources and Causes of Oil Spills

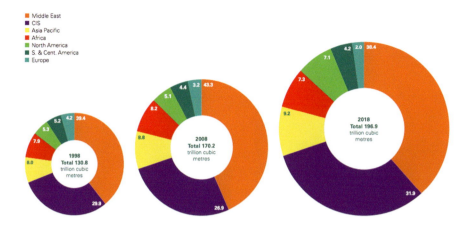

FIGURE 3.7 Percentage distribution of proved reserves of natural gas in 1998, 2008, and 2018 (BP, 2019).

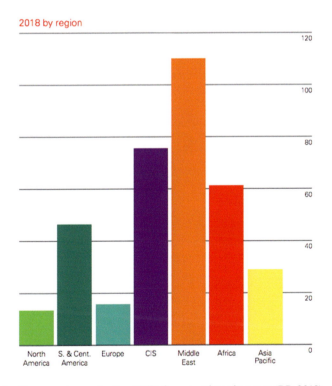

FIGURE 3.8 Reserves-to-production (R/P) for natural gas in years (BP, 2019).

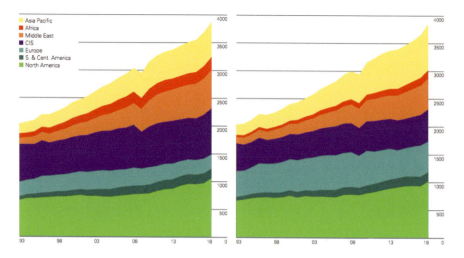

FIGURE 3.9 Natural gas production (left) and consumption (right) by region in billion cubic meters (BP, 2019).

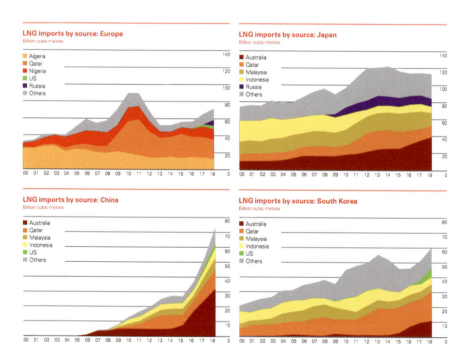

FIGURE 3.10 LNG imports by region in billion cubic meters during the 2000–2018 period (BP, 2019).

Sources and Causes of Oil Spills 55

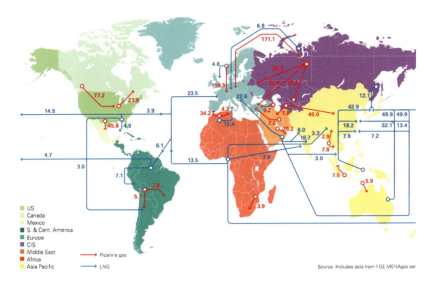

FIGURE 3.11 Worldwide major trade movements for LNG in billion cubic meters in 2018 (BP, 2019).

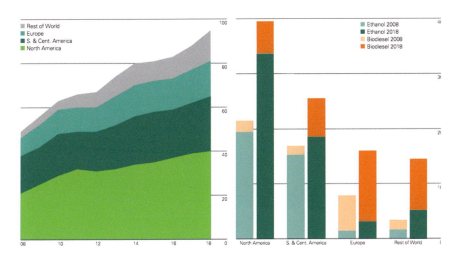

FIGURE 3.12 Worldwide biofuels production in million tons oil equivalent during 2008–2018 (BP, 2019).

3.3 OIL SPILLS CAUSED BY PETROLEUM EXPLORATION AND PRODUCTION

Approximately about 30% of global oil consumption comes from offshore production as shown in Figure 3.13, although the share of offshore production is slightly decreasing. A future scenario for global oil and gas offshore production is shown in Figure 3.14. According to the NPS scenario, offshore oil and gas production in 2040

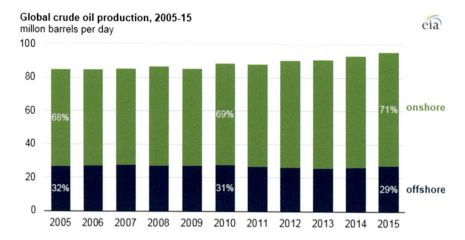

FIGURE 3.13 Share of offshore oil production during 2005–2015 (source: EIA).

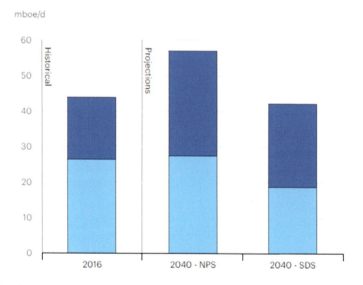

FIGURE 3.14 Offshore oil and gas production by scenario, 2016–2040. Legends: Colors: Blue: oil, dark blue: natural gas, NPS: new policies scenario, SDS: sustainable development scenario (source: IEA, 2018).

would be about 27.4 and 29.6 million barrel oil equivalent per day (mboe/d), respectively (IEA, 2018). The areas with major offshore oil and gas production include the Gulf of Mexico (US and Mexico), California (Los Angeles Basin), Arctic Ocean (Alaska), Atlantic Canada (Nova Scotia and Newfoundland), South America (Brazil and Guyana) North Sea (Norway and Scotland), Wes Africa (Angola), Caspian Sea, Persian Gulf (mainly from Iran, Qatar, Saudi Arabia, Kuwait, and UAE), Western Offshore of India, Eastern South China Sea (China, Malaysia, and Vietnam), Sarawak Field (Malaysia), Pechora Sea (Russia), and Western and North West Australia.

Sources and Causes of Oil Spills

FIGURE 3.15 Deepwater Horizon rig in the Gulf of Mexico before explosion (from Report to the President, US Government source).

A photo of an offshore platform (Deepwater Horizon) in the Gulf of Mexico before its explosion on April 20, 2010, is shown in Figure 3.15. A platform associated with pipes, risers, and the FPSO is shown in Figure 3.16 (Leffler et al., 2011). The floating production storage and offloading (FPSO) system as shown in Figure 3.17 is used extensively by oil companies for the purpose of storing oil from the oil rigs in the middle of the ocean. It is one of the best-devised systems to have developed in the oil exploration industry in marine areas (Anish, 2016). It receives reservoir fluids from subsea through risers and then separates them into crude oil, natural gas, and water through the facilities onboard.

There have been a number of major oil spill accidents that were caused by offshore operations during oil production in the sea. Nowruz oil spill in the Persian Gulf in February 1983 was caused by damage to the platform when it was struck by a tanker. The impact of the tanker, as well as waves and corrosion, was the reason the platform was toppled, releasing about 733,000 barrels of oil (1500 b/d) before it was stopped in September 1983. Two years later there was another accident at the same platform because the capping and repair were performed under fire during the war and additional 1 million barrels were released into the sea (Britannica).

The biggest oil spill caused by an offshore production facility was the Deepwater Horizon spill in the Gulf of Mexico, also known as the GOM or BP oil spill. This

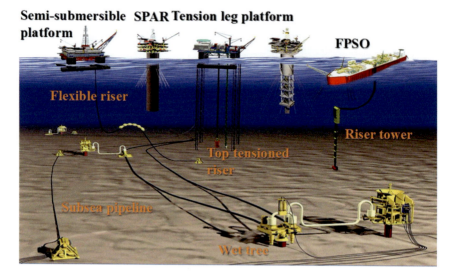

FIGURE 3.16 Platforms, offshore pipelines, and associated facilities (ASTM Manual 73).

was the largest oil spill in the Gulf of Mexico and in the United States. On April 20, 2010, an explosion as a result of a natural gas surge blasted through a well cap that was installed by the Deepwater Horizon platform. Gas ignition killed 11 people and wounded 17 and caused a flow of oil from the damaged Macondo well into water column. The well was finally capped after 87 days on July 15, after releasing between 4

FIGURE 3.17 Floating production storage and offloading (FPSO) system (www.marineinsight.com).

Sources and Causes of Oil Spills

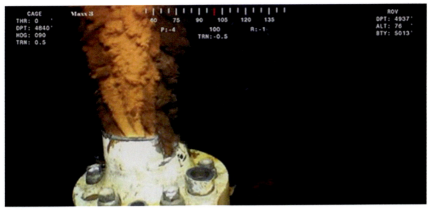

(a)

Oil tankers in New York Harbor

(b)

FIGURE 3.18 (a) Oil and hydrocarbons escaping from BP's Macondo well on June 3, 2010, during the Deepwater Horizon oil spill (source: NOAA/DOE [2010], https://www.gulfspillrestoration.noaa.gov/sites/default/files/wp-content/uploads/Chapter-2_Incident-Overview.pdf). (b) An oil tanker in New York Harbor (EIA, 2014).

and 6 million barrels of oil. The volume of oil indicated in most sources was about 4.9 million barrels. As this is the second largest oil spill in human history, with enormous economic and environmental damage, it will be discussed in the next three chapters with full detail, and its modeling is discussed in Chapters 7 and 8. Figure 3.18a shows the flow of oil from the damaged well in the Gulf of Mexico.

3.4 OIL SPILLS CAUSED BY MARINE TRANSPORTATION

There are different methods of transportation of petroleum fluids. The main factor in selecting an appropriate transportation mode is the cost: the commodity is moved from one location to another at minimum cost. The cost of transportation is mainly a function of distance, volume of fluid, and type of fluid. In summary the best method of transportation of petroleum fluids at sea is by ship and on land by pipeline. However, offshore pipelines or submarine pipelines are commonly used for short distances and especially for gas transportation. In this section we discuss the principles of these methods of transportation and their role in oil spill occurrence.

3.4.1 Oil and Gas Transportation by Tankers

Marine (or maritime) transportation is a very old and a major method of transferring goods from one location to another through waterways, rivers, seas, oceans, lakes, dams, ports, channels, canals, chokepoints, and locks. Oil is the largest commodity moved by water, which is the most cost-effective and efficient method of transportation. In the US in 2012 almost 42% of all waterborne trade was comprised of crude oil or petroleum products. Just one average-size tanker of about 30,000 tons can carry as much oil product (gasoline, diesel, and heating oil) as 1,700 tanker trucks (API, 2020).

Papavinasam reported that in 2014 more than 9,300 tankers (about 11% of total world ships) were carrying oil across the world. According to EIA, the capacity of oil tankers increased from 2,000 million barrels in the year 2000 to 4,200 million barrels in 2019. However, if condensates and petroleum products are also included, the total petroleum maritime trade is more than 60 million b/d (EIA, 2019a). An oil tanker in the waters of New York Harbor is shown in Figure 3.18b.

The size of an oil tanker is based on the amount of oil that it can carry in metric tons, and this is known as dead weight (DWT) and varies from a few to several hundred thousand DWT as shown in Figure 3.19 (EIA, 2014). The largest normal tankers are very large crude carriers (VLCC) which can carry approximately 2 million barrels of oil. The relation between the volume of oil in barrels and its weight in metric tons depends on oil density or API gravity. For a typical oil with API gravity of 33 (specific gravity of 0.86) each ton of crude is equivalent to 7.31 barrels, while for heavier oils with API gravity of 27 (specific gravity of 0.892) the conversion factor from ton to barrel is about 7 (1 ton ~ 7 bbl).

An overview of tanker size, type, length and beam (maximum width), and draft is given in Table 3.2. A typical VLCC is about 300–330 m long with a beam of 50–60 m that may be loaded or discharged within a day with 2 million barrels of oil capacity. One of the major maritime trade routes for petroleum transportation is the Strait of Hormuz and the Persian Gulf region as shown in Figure 3.20. The volume of oil transported through this route is given in Table 3.3 (EIA, 2019a). As shown in this table, about one-third of total maritime petroleum trade in the world is through the Persian Gulf. Figure 3.21 shows the amount of maritime trade for both crude oil and its products through the Strait of Hormuz.

As about 60% of world oil production moves on marine routes, for the sake of global energy security there are security chokepoints for the routes specified. Chokepoints are narrow channels along global sea routes. Some of these channels are so narrow

Sources and Causes of Oil Spills

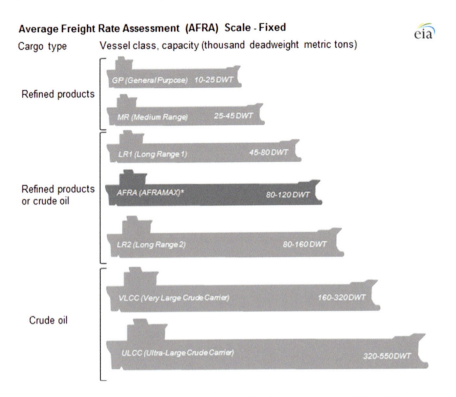

FIGURE 3.19 Various tanker sizes and their uses in oil transportation (EIA, 2014).

TABLE 3.2
Overview of Tanker Size and Type

Class	Length	Beam	Draft	Overview
Coastal tanker	205 m	29 m	16 m	Less than 50,000 dwt, mainly used for the transportation of refined products (gasoline, gasoil).
Aframax	245 m	34 m	20 m	Approximately 80,000 dwt (average freight rate assessment).
Suezmax	285 m	45 m	23 m	Between 125,000 and 180,000 dwt, originally the maximum capacity of the Suez Canal.
VLCC	330 m	55 m	28 m	Very large crude carrier. Up to around 320,000 dwt. Some can be accommodated by the expanded dimensions of the Suez Canal. The most common length is in the range of 300 to 330 meters.
ULCC	415 m	63 m	35 m	Ultra large crude carrier. Capacity exceeding 320,000 dwt. The largest tankers ever built have a deadweight of over 550,000 dwt.

Source: ASTM MNL73.

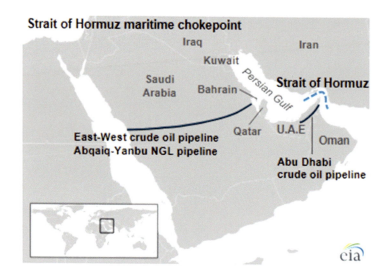

FIGURE 3.20 Persian Gulf and Strait of Hormuz (EIA, 2019a).

TABLE 3.3
Maritime Oil Trade in the Strait of Hormuz and the World

	2014	2015	2016	2017	2018
Total oil flows through the Strait of Hormuz	17.2	18.4	20.6	20.3	20.7
Crude and condensate	14.4	15.2	17.3	17.2	17.3
Petroleum products	2.8	3.2	3.3	3.1	3.3
World maritime oil trade	56.4	58.9	61.2	62.5	N/A
World total petroleum and other liquids consumption	93.9	95.9	96.9	98.5	99.9
LNG flows through the Strait of Hormuz (Tcf per year)	4.0	4.2	4.2	4.1	4.1

Source: EIA (2019).

that there are restrictions for the vessel going through them. The blockage of these chokepoints can lead to serious energy security issues in the world. Oil tankers going through chokepoints are vulnerable to pirates, terrorist attacks, political unrest, and hostilities as well as shipping accidents. Based on the volume of oil moved, the Strait of Hormuz (out of Persian Gulf) and the Strait of Malacca, linking the Indian and Pacific Oceans, are the world's most strategic chokepoints. Other major chokepoints are the Suez Canal (Mediterranean and Red Sea), the Strait of Bosporus (between the Mediterranean and Marmara Seas), the Danish Strait, and the Panama Canal linking the Atlantic and Pacific Oceans via the Sea of Caribbean. The volumes of

Sources and Causes of Oil Spills

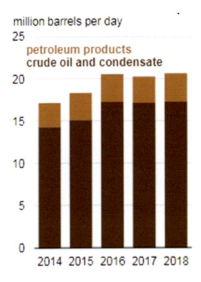

FIGURE 3.21 Flow of crude oil and petroleum products through the Strait of Hormuz in the Persian Gulf.

oil moved through these chokepoints between 2007 and 2010 are given in Table 3.4 (ASTM MNL73). The Suez Canal connects the Red Sea and Mediterranean Sea and is a critical chokepoint because of the large volume of crude oil and LNG moved from the Persian Gulf to Europe and North America. The SUMED pipeline accounts for about 8% of global LNG trade (EIA, 2019a).

One of the worst oil spill accidents known as the *SS Atlantic Empress* was caused by the collision of two oil tankers about 10 miles off the coast of Tobago in the Caribbean. A Greek oil tanker *SS Atlantic Empress* collided with the *Aegean Captain* oil tanker during a tropical rainstorm on July 19, 1979. Following the collision, both

TABLE 3.4
Volume of Crude Oil and Petroleum Products Transported through World Chokepoints, 2007–2010

Location	2007	2008	2009	2010
Bab-el-Mandab	4.6	4.5	2.9	2.7
Turkish Straits	2.7	2.7	2.8	2.9
Danish Straits	3.2	2.8	3.0	3.0
Strait of Hormuz	16.7	17.5	15.7	15.9
Panama Canal	0.7	0.7	0.8	0.7
Suez Canal and Sumed Pipeline	4.7	4.6	30.	3.1

Sources: EIA and ASTM MNL73.

vessels began spilling about 287,000 tons (more than 2 million barrels) of crude oil, causing fire on the ship and killing several crewmen (ITOPF, 2020).

Another accident occurred on 15 November, 1979, when a Romanian tanker *Independenta* collided with the Greek cargo *Evriali* at the entrance of the Bosphorous in Turkey. About 94,000 tons of Libyan oil spilled and caught fire, causing the deaths of some 42 crew members. Further explosions occurred in December causing a greater release of oil (ITOPF, 2020). A more recent example although at a smaller size occurred on February 22, 2014, when a barge collided with a towboat which caused the leak of 750 barrels of crude oil into the Mississippi River in the United States (NOAA, 2019). An oil tanker carrying 270,000 metric tons of crude oil from Kuwait to India caught fire on September 3, 2020 as a result of a boiler explosion in the engine room. The ship was the Panama-registered MT *New Diamond*, and the accident occurred off the coast of Sri Lanka, and it was confirmed that a sailor was killed although the fire was brought under control and the ship was towed with the help of the Indian Navy (https://www.thehindu.com/news/international, September 5, 2020).

When two tankers or two vessels collide with each other, the term **collision** is used. However, when a ship or tanker strikes a stationary object it is also referred to as an **allision**, although these two terms sometimes are used instead of each other. An example of an allision is the *Exxon Valdez* accident (Figure 3.22). The *Exxon Valdez* oil spill was one of the largest oil spills in the US, and occurred when the oil tanker ran hard aground on Bligh Reef on March 24, 1989, spilling 11 million gallons of crude oil (about 37,000 tons or 270,000 barrels), and was the largest oil

FIGURE 3.22 The *Exxon Valdez* oil spill was caused when a tanker ran hard aground (NOAA photo).

spill disaster in American history at that time. A storm blew in, causing oil to spread widely and polluting some 1,000 miles of coastline and killing hundreds of thousands of birds and animals (Leahy, 2019).

At the height of the BP oil spill crisis on May 25, 2010, Malaysian-registered tanker *Bunga Kelana 3* was damaged in a collision with another ship, spilling 2,000 tons of light crude oil in the Strait of Singapore, one of the most important shipping lanes in the world. The speed of response reduced the impact on the environment as was reported by Reuters (ABC, May 25, 2010). Perhaps the worst oil tanker collision occurred on January 6, 2018, when two tankers collided off the coast of China about 300 km from Shanghai. An Iranian oil tanker, the *Sanchi*, lost 117,000 tons or nearly 1 million barrels of condensate, assuming specific gravity of 0.75. The gas condensate was highly toxic and the vessel caught fire, killing all 32 crew members and sinking after a week (Leahy, 2019). More details about these oil spills are given in Chapter 4.

3.4.2 Oil and Gas Transportation by Pipelines

The transportation of crude oil, petroleum products, natural gas, LNG, and natural gas liquid (NGL) by pipeline is perhaps the most economic mode of onshore transportation (ASTM MNL58). In addition, there are offshore pipelines for the transportation of these commodities (Figure 3.23). In 2018, China completed the country's longest petroleum pipeline under the sea in the South China Sea (CCTV, 2018). SK Engineering & Construction Company (South Korea) completed an undersea pipeline in the Persian Gulf as shown in Figure 3.24 (VM, 2017).

FIGURE 3.23 The offshore pipeline is the next big thing in the pipeline development (www.pipeline-journal.net).

FIGURE 3.24 Offshore pipeline construction in the Persian Gulf (VM, 2017).

The total length of oil and gas pipeline in the world is more than 2 million km according to a report by Global Data (2019). The total length of crude oil pipeline is 379,000 km, that of petroleum products pipeline is about 267,000 km, that of natural gas pipeline is about 1.3 million km, and NGL pipelines constitute more than 92,000 km. As shown in Figure 3.25, North America has the highest oil and gas pipeline length with 41% of total pipeline length in the world. With the increase in oil and gas production and the increase in consumption in the

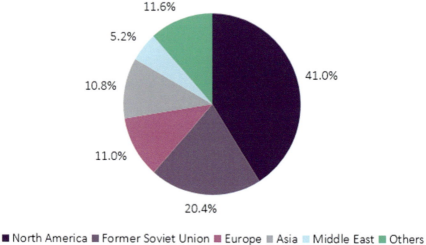

FIGURE 3.25 Share of global oil and gas pipeline length by region, Sept 2019 (source: Midstream Analytics, GlobalData Oil and Gas © GlobalData [2019]).

Sources and Causes of Oil Spills

US, there was a great American oil and natural gas pipeline boom in 2018, as discussed by Forbes (2019).

Another use of pipelines in undersea transportation is for offshore production facilities as was shown in Figure 3.16. Risers are the connection between subsea and production facilities at the surface. Materials including hydrocarbons and injection fluids are brought to the water surface from the seafloor by pipes known as risers. Risers are either rigid or flexible but insulated to withstand seafloor temperatures. Water temperature may vary from 30°C at the surface to −1°C at the seafloor (Leffler et al., 2011).

The choice between pipeline and ship is mainly a matter of distance. For short distances, transportation by pipeline is the most economical method as pipelines are easy to manufacture, have a longer life, lower maintenance costs, and better assurance of continuous delivery. For natural gas transportation, the use of pipeline is mainly a factor of capacity of gas delivery and distance. For radius of greater than 2000 km, marine transport should be used instead of pipeline. For the case of long distances, the best option would be to convert natural gas into liquid form known as LNG, or to convert to liquid through processes such as Firscher and Tropsch known as gas to liquid (GTL). Compressed natural gas (CNG), used for short distances, requires pressurizing the gas through pump stations. In the LNG process the volume of gas can be reduced significantly by a factor of 1:600, and the LNG option dominates the market for the marine transportation of natural gas (ASTM MNL58).

One of the biggest sources of pipeline failure and causes of leaks is due to internal corrosion of the pipe. For example, on January 8, 1999, a leak in a 12-inch pipe at an industry tank farm in Benavides County, Texas, spilled more than 10,000 barrels of crude oil, and the cause was internal corrosion. Later the EPA announced that the responsible company paid the largest civil fine for 300 oil spills for its pipelines and oil facilities in six states, and it was requested that the company improve its leak-prevention program and help environmental projects (EPA, 2000). In 1994 the Kolva River oil spill was the largest oil spill caused by corroded pipeline in the Russian Arctic. About 2 million barrels of oil were released, mainly into the landscape but some into the river as well.

In January 1969, nearly 3 million gallons of oil spewed from a drilling rig near the coast of Santa Barbara, California. The pipe blowout cracked the seafloor and the oil plume killed thousands of seabirds. About 35 miles of coastline were coated with oil up to 6 inches thick, and about 800 square miles of ocean were affected (CNN, 2015). On January 18, 2000, a pipeline owned by Petrobras ruptured and spewed about 8,200 barrels of heavy oil into Guanabara Bay outside Rio de Janeiro in Brazil (US EPA, 2001). Another example in which pipeline damage caused an oil spill accident is Statfjord Oil Spill on December 12, 2007, during the loading of a tanker off the coast of Norway. More than 21,000 barrels of oil were spilled into the North Sea. This was the second largest oil spill in Norway after the Ekofisk oil spill in 1977. The cause of the accident as shown in Figure 3.26 was a break in the hose between the seabed and the tanker connection. The spill covered an area of 23 km^2 on the same day, but after one day the spill was 10 km long and 5 km wide with a thickness of 100 microns or 0.1 mm (EMSA, 2013).

FIGURE 3.26 Damaged hose that caused second largest oil spill in Norway (source: EMSA, 2013).

As this book is being written and on June 12, 2020, the Globe and Mail reported that there was an oil spill released from the Canadian government-owned Trans Mountain pipeline near Vancouver in British Columbia. The company delivers 300,000 barrels of crude oil every day through a 1150-km pipeline from Alberta and British Columbia with 111 km in the state of Washington in the US. The company said the spill was contained and the crew immediately began the cleanup operation; however, the amount of oil spilled is not yet reported.

FIGURE 3.27 An oil refinery is seen near an oil slick along the coast of Refugio State Beach in Goleta, California on May 19, 2015 (*The Atlantic*, 2015).

3.5 OIL SPILLS CAUSED BY REFINERIES

Many refineries around the world are built and located near coastline and seas or oceans in order to receive crude oil or to deliver their products to the market by marine transportation. One of these refineries, shown in Figure 3.27, is located near Santa Barbara, California, which experienced an oil spill accident in 2015. The cause of the accident was the rupture of the onshore pipeline delivering crude oil to the refinery.

The Hazardous Materials Safety Administration says preliminary findings indicate the pipeline was "experiencing active external corrosion" and that metal loss amounted to 45% of the pipeline wall where the pipe broke. Other sections of the pipeline showed almost three-quarters of the pipe corroded, according to federal investigators. The pipeline ruptured and spilled about 105,000 gallons of crude oil along the California coastline near Santa Barbara. More than 20,000 gallons of the oil was spilled into the Pacific Ocean 20 miles west of Santa Barbara (NPR, 2015). The oil slick spread at least four miles along Refugio State Beach by sunset as shown in Figures 3.28 and 3.29.

3.6 OIL SPILLS CAUSED BY NATURAL SEEPS

According to the National Oceanic and Atmospheric Administration (NOAA), an oil seep is a natural leak of crude oil and gas that migrates up through the sea floor and ocean depths. For example, a natural tar seep offshore of Santa Barbara from the US Geological Survey is shown in Figure 3.30 (NOAA, 2020).

FIGURE 3.28 An oil slick spreads along the coast of Refugio State Beach in Goleta, California, after an oil pipeline ruptured on May 19, 2015 (*The Atlantic*, 2015).

FIGURE 3.29 Spilled oil covers Refugio State Beach and other beaches north of Goleta, California (https://nrm.dfg.ca.gov/FileHandler.ashx?DocumentID=178492&inline).

FIGURE 3.30 A natural tar seep in Santa Barbara offshore (NOAA, 2020).

Leaks from seeps occur wherever there are fractures in the seafloor, similar to the way a freshwater spring brings water to the surface. When there is an oil spill it is important to know the type of oil, if it is coming from a production platform or from a seep, as these two oils are different. In addition, the weather conditions, temperature, wind, tides, and currents affect the movement of oil after it is generated from

Sources and Causes of Oil Spills

FIGURE 3.31 Naturally occurring oil spill. Locations of oil seeps offshore remain consistent over time (NOAA, 2019).

a seep. One major difference between oil released from a seep and that from human error is the speed at which the oil spreads over water. Natural oil spills generated from seeps move very slowly, while oil spills from a tanker collision quickly cover vast areas as they move fast and in large quantities. However, both types of spills damage the environment in the same way.

Natural oil seeps are perhaps the largest source of oil entering the world ocean and roughly account for half of oil released into the sea annually. A study by the National Research Council indicates that on average approximately 16,000 tons of naturally occurring oil spills enter North American waters each year. There is a rough estimate that the natural seeps off the coast of Santa Barbara in California have effectively generated an oil spill of about 5 million gallons of oil leaked from the seafloor each year for the last several hundred thousand years. An oil slick from natural seeps is shown in Figure 3.31. Although the rate of oil released may vary with time, the locations are consistent and predictable. Some slicks are large enough to be seen from satellites. Oil spill from seeps is sticky and thick, dark like used motor oils (NOAA, 2019). Another study carried out by scientists from the Woods Hole Oceanographic Institution (WHOI) and the University of California Santa Barbara (UCSB) shows that about 35,000 years ago a series of apparent undersea volcanoes deposited a massive flow of petroleum 10 miles offshore and the deposit hardened with time into domes that were discovered recently. These domes are 700 feet deep in the waters off the coastal resort of Santa Barbara (WHOI, 2010).

3.7 OIL SPILLS CAUSED BY WAR, SABOTAGE, AND OTHER ACTIONS

War and terrorism have both direct and considerable impacts on the environment and financial damage to people affected by the incident. Oil and gas installations such as production sites, refineries, transportation, and storage facilities are the economic backbone of oil-producing countries and are major war and sabotage targets. This is particularly the case for oil-rich countries involved in political conflicts such as in the Middle East and North Africa regions. During the eight-year war between Iran and Iraq in 1980–1988 many oil installations in the Persian Gulf were destroyed by air or naval operations. For example in 1984 alone about 56 attacks were conducted against oil tankers in the Persian Gulf. More than 60% of ships attacked during the war were oil tankers, and in total about a quarter of all oil tankers operating in the region were fully destroyed (Figure 3.32). The tanker war in the region continued for four years, and Iran's oil export was reduced by half and overall shipping in the Gulf was reduced by 25%. Most recently a tanker in the Gulf of Oman was hit in 2019 as shown in Figure 3.33. In October 1987, the Sassan and Rostam platforms, 30 miles apart, east of Qatar and 60 miles south of the Iranian coast, were attacked and caught fire (Los Angeles Times, 1987). Figure 3.34 shows one of these platforms on fire after the attack, and the locations of these platforms along with other attacks during the 1986–1988 period are shown in Figure 3.35. A good review of the tanker war during 1987–1988 in the Persian Gulf was given by the Washington Institute (2009) and Mercogliano (2019).

There was a second war in the Persian Gulf Region in 1990–1991 during the Iraqi invasion of Kuwait and the follow-up expulsion of Iraq from Kuwait by US forces.

Iraqi forces, in a deliberate action, set about 700 wells on fire and dumped millions of barrels of crude oil into the Persian Gulf to stop US Marines from landing. This sabotage, along with damage to oil tankers, oil refineries, and Iraqi oil terminals, created the worst oil spill in human history, with about 5 million barrels (initial estimate was 6 to 8 million) of oil on the water surface (Figure 3.36). The oil spill covered some 770 km of coastline from Kuwait to Saudi Arabia and UAE, and about 1550 square km of sea surface with a thickness of about 13 cm (Devastating

FIGURE 3.32 An oil tanker in the Persian Gulf attacked during the 1987–1988 naval war (source: https://historica.fandom.com/wiki/Tanker_War?file=Tanker_War.jpg).

Sources and Causes of Oil Spills 73

FIGURE 3.33 Attack on an oil tanker in the Gulf of Oman on June 13, 2019 (source: Sal Mercogliano, 2019).

Disasters, 1991). More details about this oil spill are given in the next chapter. While writing this chapter, major news agencies reported today (May 11, 2020) that a missile was fired by mistake in the Gulf of Oman (Figure 3.37), destroying a ship with an initial estimate of 19 killed and 15 wounded.

As another example, during the Balkan War in 1999, NATO bombing raids damaged petrochemical plants and refineries, releasing oil and hazardous materials into above and below ground waters in Serbia. Extensive oil slicks occurred in the

FIGURE 3.34 An oil platform attacked in October 1987 in the Persian Gulf (source: Washington Institute).

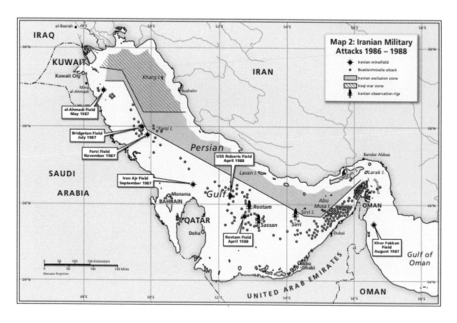

FIGURE 3.35 Map of Persian Gulf and attacks during 1986–1988 (source: Washington Institute).

Danube, and finally these pollutants affected rivers in neighboring countries such as Romania and Bulgaria (Mannion, 2003).

Oil tankers are also attacked by terrorists as it is easy for them to attack ships due to the lack of security at sea. For example, on October 6, 2002, the supertanker *Limburg* was carrying about 400,000 barrels of crude oil when it was attacked by Al-Qaida terrorists at the Horn of Africa. It created an oil spill of about 90,000 barrels into water in the Gulf of Aden and significant damage to the ship as well as one death. In the same year the Moroccan government arrested several members of al-Qaida, accusing them of plotting to attack British ships in the Strait of Gibraltar (Guardian, 2002). A good review of terrorist attacks on oil installations and tankers is given by Steinhäusler et al. (2008).

Another type of accident that causes oil spills is due to wrecks and the grounding of oil tankers, usually referred as **wreck spills**. An example of such oil spills is the accident that happened on January 5, 1993, in Shetland, UK. Following engine failure, the *Braer* tanker ran aground in severe weather conditions. The ship was carrying more than 86,000 tons (about 600,000 barrels) of Norwegian crude oil, and over 12 days the entire cargo was lost as heavy seas broke the ship apart. The oil was exceptionally light, and combined with strong wind and waves the oil was vaporized or dispersed into water quickly (ITOPF, 1995).

In another accident on October 5, 2011, a Liberian ship called the *Rena* grounded on Astrolabe Reef 14 miles offshore from Tauranga Harbor on New Zealand's North Island. It was carrying more than 1,700 tons of oil and 200 tons of diesel fuel. A

Sources and Causes of Oil Spills 75

FIGURE 3.36 During the 1991 war Iraqi forces set fire to hundreds of oil wells in Kuwait, creating the largest oil spill (more than 5 million barrels) in human history (commons.wikimedia.org).

FIGURE 3.37 A missile was fired in the Gulf of Oman on May 11, 2020 (CBC, 2020).

FIGURE 3.38 Oil spill after Hurricane Rita in November 2005 (photo: NOAA).

cracked hull and rough seas caused the dislodging of containers, killing 1,000 birds in the first week after the accident (Atlantic, 2011).

In an older accident on March 16, 1978, *Amoco Cadiz*, a VLCC, ran aground off the coast of Brittany in France. The amount of oil released was more than 1.6 million barrels of light oil, and within a month it had contaminated 320 km of the coastline. The ship was damaged by a large wave in the English Channel. With advances in navigation technology the number of such major accidents has been reduced significantly in recent years. In November 2005, a double-hulled barge carrying about 5 million gallons of oil sank due to Hurricane Rita as shown in Figure 3.38.

3.8 SUMMARY

The purpose of this chapter was to present the sources, causes, and occurrence of oil spills. The chapter began with some statistical data on the existing proved reserves of oil and gas and their production rates, followed by identifying producing and consuming countries as well as the lifetime of oil and gas. Methods of transportation of oil and gas especially by tankers and pipelines were discussed. Major causes of oil spills due to offshore production activity, collision, grounding, equipment failure, wreck, war, sabotage, and terrorism were discussed with some examples for each case.

The amount of oil spilled mainly depends on the cause of the oil spill as shown in Figure 3.39. Spills can be divided into three groups of small (<7 tons), medium (7–700 tons), and large (>700 tons). Collision and grounding are the causes of more

Sources and Causes of Oil Spills

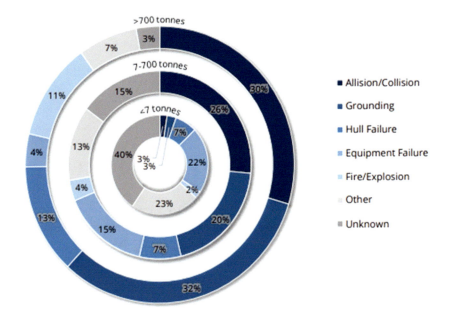

FIGURE 3.39 Cause of oil spills (ITOPF, 2019).

than 60% of large oil spills, while small spills occur mainly due to other or unspecified causes.

REFERENCES

Anish. 2016. *Types of ships, what is FPSO (Floating Production Storage and Offloading) system?* July 21. Taken from the following link on May 6, 2020. https://www.marine-insight.com/types-of-ships/what-is-fpso-floating-production-storage-and-offloading-system/

API. 2020. *Maritime safety and efficiency.* American Petroleum Institute, Wahsington, DC. Taken from the following link on May 5, 2020. https://www.api.org/oil-and-natural-gas/wells-to-consumer/transporting-oil-natural-gas/oil-tankers/maritime-safety-and-efficiency

Atlantic. 2011. *Oil spill disaster on New Zealand shoreline, by Alan Taylor.* October 14. https://www.theatlantic.com/photo/2011/10/oil-spill-disaster-on-new-zealand-shoreline/100169/

Atlantic. 2014. *Remembering the Exxon Valdez oil spill by Allan Taylor.* March 24. Taken from the following link on May 7, 2020. https://www.theatlantic.com/photo/2014/03/remembering-the-exxon-valdez-oil-spill/100703/

Atlantic. 2015. *An oil spill fouls the California coastline by Alan Taylor.* May 20. Taken from the following link on May 8, 2020. https://www.theatlantic.com/photo/2015/05/oil-spill-fouls-california-coastline/393731/

BP. 2019. *BP statistical review of world energy*, 58th Edition. https://www.bp.com/content/dam/bp/business-sites/en/global/corporate/pdfs/energy-economics/statistical-review/bp-stats-review-2019-full-report.pdf

Britannica. *Encylopedia, biggest oil spills in history.* Taken from the following link on May 11, 2020. https://www.britannica.com/list/9-of-the-biggest-oil-spills-in-history

CBC. 2020. *Canadian broadcasting corporation.* May 11. https://www.cbc.ca/news/world/iran-navy-missile-1.5564120

CCTV. 2018. *China completes longest undersea pipelines in South China Sea.* Post on CCTV News Agency on February 18.

CNBC. 2020. Taken from the following link on May 4, 2020. https://www.cnbc.com/2020/05/05/oil-markets-coronavirus-easing-lockdowns-in-focus.html

CNN. 2015. *Santa Barbara oil spill: Crude flowed 'well below' capacity in ruptured pipe, CNN report on May 22, 2015.* Taken from the following link on May 10, 2020. https://www.cnn.com/2015/05/21/us/california-oil-spill/index.html

Devastating Disasters. 1991. *Gulf war oil spill.* January 21. https://devastatingdisasters.com/gulf-war-oil-spill-1991/

EIA. 2013. *Today in energy, Energy Information Administration (EIA).* February 1. https://www.eia.gov/todayinenergy/detail.php?id=9811

EIA. 2014. *Today in energy, oil tanker sizes range from general purpose to ultra-large crude carries on AFRA scale, contributor: T. Mason Hamilton, US energy information administration.* September 14. Taken from following link on May 5, 2020. https://www.eia.gov/todayinenergy/detail.php?id=17991

EIA. 2019a. *Today in energy, The Suez canal and SUMED pipeline are critical chokepoints for oil and natural gas trade.* July 23. https://www.eia.gov/todayinenergy/detail.php?id=40152

EIA. 2019b. *Today in energy, US energy information administration.* June 20. https://www.eia.gov/todayinenergy/detail.php?id=39932

EMSA. 2013. *European maritime safety agency.* As adopted by EMSA´s Administrative Board at its 37th Meeting held in Lisbon, Portugal, on November 13–14, 2013. p. 38. file:///C:/Users/User/Downloads/action-plan-for-response-to-marine-pollution-from-oil-and-gas-installations%20(1).pdf

EPA. 2000. *Koch industries to pay record fine for oil spills in six states. US environmental protection agency.* Release Date: January 13. Taken from link below on May 8, 2020. https://archive.epa.gov/epapages/newsroom_archive/newsreleases/981d17e5ab07246f8525686500621079.html

EPA. 2001. *US environmental protection agency archive document on petrobras accidents.* https://archive.epa.gov/emergencies/content/fss/web/pdf/souzapaper.pdf

Forbes, G.L. 1994. The Braer oil spill incident—Shetland, January 1993. *International Journal of Environmental Health Research,* 4, pp. 48–59. https://www.tandfonline.com/doi/abs/10.1080/09603129409356797

Forbes, G.L. 2019. *The Great American oil and natural gas pipeline boom by judge clemente.* Taken from the following link on August 4, 2019. https://www.forbes.com/sites/judeclemente/2019/08/06/the-great-american-oil-and-natural-gas-pipeline-boom/#190f14be1512

GlobalData. 2019. *North America has the highest oil and gas pipeline length globally.* December 4. Taken from the following link on May 8, 2020. https://www.offshore-technology.com/comment/north-america-has-the-highest-oil-and-gas-pipeline-length-globally/

Globe and Mail. 2018. *Crews start cleaning up California coastline after broken pipe spills oil into sea by Christopher Weber, Associated Press.* May 15. Taken from the following link on May 8, 2020. https://www.theglobeandmail.com/news/world/about-21000-gallons-of-oil-spilled-into-ocean-from-california-pipeline/article24507978/

Guardian. 2002. *Al-Qaida suspected in tanker explosion, reported in the Guardian on October 7, 2002.* https://www.theguardian.com/world/2002/oct/07/alqaida.france

Sources and Causes of Oil Spills 79

IEA. 2018. *Offshore energy outlook, international energy agency, taken from link below on May 6, 2020.* https://www.iea.org/reports/offshore-energy-outlook-2018

ITOPF. 1995. *Braer, UK 1993.* January 5. https://www.itopf.org/in-action/case-studies/case-study/braer-uk-1993/

ITOPF. 2019. *Oil tanker spill statistics 2018.* January. http://www.itopf.org/fileadmin/data/Documents/Company_Lit/Oil_Spill_Stats_2018.pdf

ITOPF. 2020. International Tanker Owners Pollution Federation Limited, ITOPF, Ltd, London. Taken from the following link on May 5, 2020. www.itopf.org

Leahy, S. 2019. Exxon Valdez changed the oil industry forever—but new threats emerge. *National Geographic.* March 22. Taken from the following link on May 7, 2020. https://www.nationalgeographic.com/environment/2019/03/oil-spills-30-years-after-exxon-valdez/

Leffler, W.L., R. Pattarozzi and G. Sterling. 2011. *Deepwater petroleum exploration and production.* PennWell Corporation, Tulsa.

Los Angeles Times. 1987. U.S. Destroys 2 Iranian Oil Platforms in Gulf, by Associated Press, October 19, 1987.

Mannion. 2003. *The environmental impact of war & terrorism.* Geographical Paper No. 169, June. The University of Reading, Whiteknights. https://www.reading.ac.uk/web/files/geographyandenvironmentalscience/GP169.pdf

Mercogliano, Sal. 2019. *The tanker war, part II: A historical perspective, GCaptain.* June 14. https://gcaptain.com/the-tanker-war-part-ii-a-historical-perspective/

NOAA. 2019. *What are natural oil seeps? National Oceanic and Atmospheric Administration (NOAA), US department of commerce.* Updated July 25, 2019. Taken from the following link on July 25, 2020 https://response.restoration.noaa.gov/oil-and-chemical-spills/oil-spills/resources/what-are-natural-oil-seeps.html

NOAA. 2020. *What is an oil seep? National Ocean Service, National Oceanic and Atmospheric Administration, US department of commerce.* Updated April 4, 2020. Taken from the following link on May 10, 2020. https://oceanservice.noaa.gov/facts/oilseep.html

NPR. 2015. *Burst oil pipeline in California severely corroded, investigators say, by Bryan Naylor.* June 4. Taken from the following link on May 9, 2020. https://www.npr.org/sections/thetwo-way/2015/06/04/411989248/burst-oil-pipeline-in-california-severely-corroded-investigators-say

OECD, Glossary of Statistical Terms. 2001. *Taken from: Glossary of environment statistics, studies in methods, series F, No. 67.* United Nations, New York, 1997. https://stats.oecd.org/glossary/detail.asp?ID=1902

Papavinasam, S.. 2014. *Corrosion control in the oil and gas industry.* Elsevier, Amsterdam.

Riazi, M.R., Ed. 2016. *"Exploration and Production of Petroleum and Natural Gas," ASTM manual (MNL) 73.* ASTM International, USA, pp. 507–529, Chapter 18, Petroleum and Natural Gas Transportation and Storage, by W. Zou, J. Lu, R. Kumar, X. Liu, and A.A. Gupta.

Riazi, M.R., S. Eser, S.S. Agrawal and J.L. Peña-Díez. Eds. 2013. *Petroleum refining and natural gas processing.* ASTM International. Manual (MNL) 58, 2013. Chapter 22: Transportation of Crude Oil, Natural Gas, and Petroleum Products by Luis Ayala.

Steinhäusler, F., P. Furthner, W. Heidegger, S. Rydell and L. Zaitseva. 2008. *Security risks to the oil and gas industry: Terrorist capabilities strategic insights, volume VII, issue 1.* Strategic Insights, the Center for Contemporary Conflict at the Naval Postgraduate School in Monterey, California. February. file:///C:/Users/User/Downloads/483063.pdf

VM. 2017. Valentime Maritime Gulf L.L.C., Abu Dhabi, October 17, http://www.vmgulf.com/.

Washington Institute. 2009. *Gulf of conflict, a history of US-Iranian confrontation at sea, by David Crist*. The Washington Institute for Near East Policy. June. Taken from the following link on May 10, 2020. https://www.washingtoninstitute.org/uploads/Documents/pubs/PolicyFocus95.pdf

WHOI. 2010. *WHOI scientists find ancient asphalt domes off California coast*. Taken from the following link on April 25, 2010. https://www.whoi.edu/press-room/news-release/whoi-scientists-find-ancient-asphalt-domes-off-california-coast/

4 Major Oil Spills
Occurrence and Their Impacts

ACRONYMS

AMSA	Australian Maritime and Safety Authority
AP	Associated Press
BP	British Petroleum
BOP	Blowout preventer
CIS	The Commonwealth of Independent States (from former Soviet Union)
DWH	Deepwater Horizon
DWT	Deadweight tonnage
EIA	Energy Information Administration (part of US Department of Energy, Washington, DC)
EPA	US Environmental Protection Agency
GOM	Gulf of Mexico
GOR	Gas-to-oil ratio
IEA	International Energy Agency (based in Paris, France)
ITOPF	International Tanker Owners Pollution Federation
LNG	Liquefied natural gas
Mb/d	Million barrels per day (1 b/d average oil is equivalent to 50 metric ton/year)
Mbbl	Million barrels of oil (~159,000 m³)
NOAA	National Oceanic and Atmospheric Administration
NCAR	National Center for Atmospheric Research
OPA	Oil Pollution Act
SAR	Synthetic-aperture radar
VLCC	Very large crude carrier

4.1 INTRODUCTION

As discussed in Chapter 3, the cause and occurrence of oil spills are usually due to offshore production activity or during the marine transportation of crude oil and its products as well as LNG, mainly by oil tankers. While offshore production has been the cause of some of the largest oil spills in history, tanker accidents have created a large number of oil spills as shown in Figure 4.1 (ITOPF, 2020). In this figure the

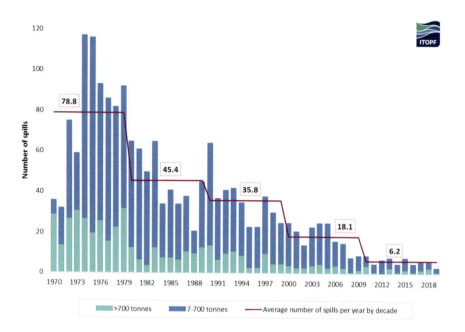

FIGURE 4.1 Number of oil spills by oil tankers (>7 tons) from 1970 to 2019 (ITOPF, 2020).

number of spills caused by oil tankers alone, greater than 7 tons (>50 barrels), from 1970 to 2019 is shown. As shown in this graph, the number of oil spill accidents has decreased in recent decades. For example, since 2010 this number has reduced to about 5 spills per year while in the 1970s this number was about 50 spills per year. A map of the locations of spills greater than 7 tons since 1989 is shown in Figure 4.2. The largest spill from an oil tanker was the *ABT Summer* off the coast of Angola in West Africa which occurred in May 1991. About 2 million barrels of heavy oil were released into the sea covering an area of about 80 square miles which was caused by fire and explosion in the ship. However, its environmental impact was not significant as it occurred 900 miles from the coast.

Since 1970 there have been more than 10,000 incidents of which the majority of spills were less than 7 tons. Figure 4.2 shows about 613 tanker accidents reported from 1989 to 2018 with spills greater than 50 barrels. However, the larger spills of greater than 700 tons (5000 barrels) reduced from an average of nine per year to two. These data collected by ITOPF (2020) should be taken with caution as some data have been revised from the date of accident and various sources reported different volumes of oil spilled.

In 1997 an accident near Sharjah in the United Arab Emirates split some 40,000 barrels of diesel oil, causing the shutdown of a 20-million-gallons-a-day desalination plant in Sharjah which supplies drinking water to some 500,000 people in the region (Arab Times, 1998). Two more accidents in the same area occurred when tankers carrying illegal Iraqi fuel and gas oils were interrupted due to UN sanctions imposed on Iraq (ASCE, 1996). In some areas, however, the number of incidents has been on the rise. For example, Nguyen (2018) reviews the accidents in South East Asia and

Major Oil Spills

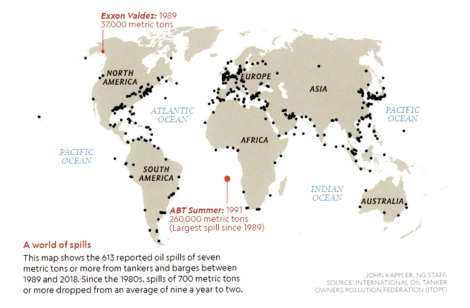

FIGURE 4.2 Tanker oils spills since 1989 (ITOPF, 2019).

Vietnam waters. Since 1993, statistics in Vietnam show an average of four spills per year which have been caused by tanker collisions.

The largest oil spill in human history occurred in the Persian Gulf in January 1991 during the Iraqi invasion of Kuwait, which spewed between 6 and 8 million barrels (about 1 million m^3) of oil. The second largest oil spill occurred in April 2020 in the Gulf of Mexico after the Deepwater Horizon accident and blowout of the Macondo well which spewed almost 5 million barrels of oil into the sea in the largest oil spill in US history. These two oil spills are discussed in more detail in this chapter.

In Europe the worst pollution disaster was the 1991 sinking of the *Haven*, an oil tanker loaded with 144,000 tons of crude oil, releasing 50,000 tons of oil into the sea, killing five people; for the next 12 years the Mediterranean coast of Italy was polluted, especially around Genoa. Examples of some other smaller oil spills are Cornwall, England (1967, 38 million gallons off the coast of the Scilly Islands), North Sea (1977, 81 million gallons due to the blowout of a well in the Ekofisk oil field of Norway), Portsall, France (1978, 68 million gallons), Gulf of Mexico (1979, 140 million gallons), Persian Gulf, Iran (1983, 80 million gallons), Canary Islands (1989, 20 million gallons of crude oil spilled in the Atlantic Ocean forming a 100 square-mile oil slick), Uzbekistan (1992, 88 million gallons), the *Hebei Spirit* oil spill, South Korea (December 2007, more than 3 million gallons), and the *Erika* oil spill in Brittany, France (December 1999, 20,000 tons or more than 6 million gallons). Another area with a high rate of oil spill occurrence is the Niger Delta where an average of 240,000 barrels of oil are spilled every year as reported in the *Nigerian Medical Journal* (Ordinioha and Brisibe, 2012). In recent years, due to advances in logistics and tanker hulls, oil spill accidents have slowed down (Riazi, 2010).

4.2 OIL SPILL DETECTION

If the volume of oil released into the sea is not known, there are other methods with which the amount of oil can be estimated from the surface area and its thickness. Vogt and Tarchi (2004) have shown how the SAR satellite system can be used to monitor hundreds of small oil spills which occur every year in the Mediterranean Sea. The detection and monitoring of oil spills have been discussed by Riazi (2010).

A geographic information system (GIS) is a computer monitoring system capable of storing, analyzing, and sharing geographical information which can be used to detect oil spills. Usually airplanes and satellites can be used to detect and monitor oil spills. Oil spill detection by synthetic-aperture radar (SAR) is based on the dampening effect oil has on seawater surface waves. SAR image classification can show three different classes for an oil spill: the spill area in the center surrounded by a high-pollution area and the outer layers of a low-pollution area (Mansor et al., 2006). At very low winds, no SAR signal is received from the sea, thus no slicks can be seen (Automatic Oil Spill Detection, 1999). The unpolluted water surface shows higher surface roughness than the oil slick which translates into increased backscattered signal (Vogt and Tarchi, 2004).

Remote sensing devices that are available for oil spill detection include infrared photography, thermal infrared imaging, airborne laser fluorosensors, airborne and spaceborne optical sensors, and airborne and spaceborne synthetic-aperture radar. Generally remote sensing devices can help to identify minor spills before they cause widespread damage. Remote sensors should be operational both day and night. Researchers indicate that a combination of airplanes and satellites equipped with SAR sensors is the most effective and valuable tool to identify oil spills (Sykas, 2020). The steps to detect an oil spill by SAR include radar image, pre-processing, and post-processing to identify the type and thickness of the spill. Figure 4.3 shows an SAR image classification showing three different classes for an oil spill: high- and low-pollution areas as well as the oil spill.

In November of 2002 off the coast of Galicia in Spain an oil spill of 11,000 tons occurred as reported by Montero et al. (2006). The spill's movement was monitored by overflights, and its path was forecasted with different models. Several helicopters and ships were used to monitor the spill with the help of some volunteers and fishermen. Information obtained from overflights was entered in a GIS to obtain a picture of the situation. The track of the oil spill is obtained by considering that its speed is about 3% of the surface wind module with the same direction (Riazi, 2010).

The aircrafts used for monitoring an oil spill must have: (1) good downward visibility, (2) good radios for direct communications with vessels or ground personnel, and (3) a global (geographic) positioning system (GPS) as discussed in the *Oil Spill Monitoring Handbook* (AMSA, 2003). By knowing the speed of the aircraft and the time taken to fly over the length and width of the spill, the length and size can be determined (note: 1 knot = 0.5 m per second or 1.8 km per hour). From observing the color of the slick, the approximate thickness and volume per square km can be determined according to data given in Table 4.1. In this table the approximate % covered

Major Oil Spills

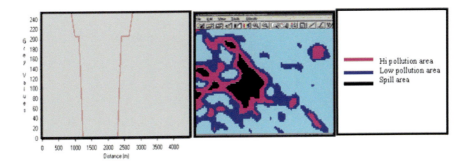

FIGURE 4.3 SAR image classification showing three different classes for oil spills (source: Mansor et al., 2006).

TABLE 4.1
Visual Method for Approximate Estimate of Thickness and Percent Coverage Area of an Oil Spill

Description/ Colour	Thickness (mm)	Volume (m^3/sq km)
Silvery sheen	0.0001	0.1
Bright bands of rainbow colour	0.0003	0.3
Dull colours seen	0.001	1.0
Yellowish brown slick	0.01	10
Light brown or black slick	0.1	100
Thick dark brown or black slick	1.0	1,000

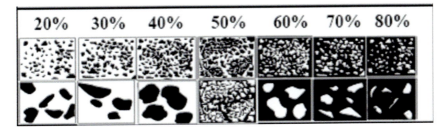

Source: AMSA (2003).

by the slick is also given based on the observation of the slick. The monitoring of marine and shoreline environments is discussed in detail by ASMA (2003).

4.3 MAJOR ACCIDENTS

A list of 13 oil spill accidents with a size of more than 1 Mbbl (42 million gallons) during the past 4 decades is given in Table 4.2. As the data show in this table oil spills caused by offshore exploration and production constitute the largest spills

TABLE 4.2
Summary of Oil Spills Greater than One Million Barrels since 1980

Order	Name	Date	Location	Cause	Approximate Amount of Oil Released	
					Million Gallons	Million Barrels (Mbbl)
1	Gulf War	January 1991	Persian Gulf, Middle East	Iraq-Kuwait War	252–336	5–8

This was the largest oil spill in human history, occurring during the Kuwait-Iraq war as Iraqi soldiers set oil production wells on fire and released oil from tankers, storage tanks, and refineries to prevent US forces from landing on Kuwaiti shores. Although some sources gave an estimate of up to 11 million barrels, a more realistic number is between 6 and 8 million barrels of oil released to the Persian Gulf. Part of the oil was washed to the shores of Kuwait and Saudi Arabia and still exists today. Further details are given Section 4.5.

| 2 | Deepwater Horizon | April–July 2010 | Gulf of Mexico, US | Wellhead blowout | 165–210 | 4–5 |

As indicated in the introduction, this was the largest oil spill accident in US history and the second largest in human history. The cause was similar to the Ixtoc production well blowout (due to poor maintenance) which occurred on 20 April 2010, and oil was released into the water near the shore of Louisiana. About 5 million barrels of oil were released within 3 months, causing enormous economic and environmental damage. As this is the largest oil spill in the US and was fully covered in the media, a large section of this chapter is devoted to this incident and its consequences as given in Section 4.6. This incident and its consequences are also covered in Chapters 5 and 6.

| 3 | Ixtoc I | June 1979–March 1980 | Bay of Campeche, Gulf of Mexico | Exploratory well blowout | 140–150 | 3.3–3.5 |

An explosion in 1979 at the Ixtoc I platform, located 80 km (50 miles) off Mexico's Gulf Coast, caused the largest ever peacetime oil spill by that time. It took more than 9 months for the Mexican's state oil company to stop the leak; during this period more than 3.3–3.5 million barrels of oil gushed into the sea. The main difference between this accident and the BP accident in 2010 is that the Ixtoc was drilling at a depth of 45 m, whilst Deepwater blew at 1500 m below the sea (BBC, 2010).

(Continued)

TABLE 4.2 (CONTINUED)

Summary of Oil Spills Greater than One Million Barrels since 1980

Order	Name	Date	Location	Cause	Approximate Amount of Oil Released	
					Million Gallons	**Million Barrels (Mbbl)**
4	Fergana Valley	March 1992	Uzbekistan	Oil well blowout	>88	>2

This oil spill occurred in a valley in Uzbekistan as a result of a blowout at a well being drilled and was the worst oil spill in the history of Central Asia. Oil was being released at a rate of 35,000 to 150,000 bbl/d. A total of 2 million barrels was released (Global Energy News, 1992). As this oil was not released in a lake or sea, some sources do not consider this as a typical oil spill. However, some of the methods for treating oil spills in water for estimation of rate of evaporation of oil can be applied to land oil spills as well.

| 5 | Nowruz oil field | February–September 1983 | Persian Gulf/Iran | Iraq-Iran war | >80 | >1.9 |

This oil spill was caused by the collision of a ship with a rig in Nowruz oil field in the Persian Gulf during the Iraq-Iran war. Altogether and with two oil spills, a total of about 2 million barrels of oil spewed into the Gulf water from September 1983 to March 1985 (Cedre, 2010).

| 6 | *ABT Summer* | May 1991 | Off the coast of Angola | A blast on board | ~80 | ~1.9 |

Liberian oil tanker *ABT* exploded due to fire on board and the entire crude oil spilled into the South Atlantic Ocean. The blast was due to corroded tank, and the tanker burned for three days before sinking. The oil slick spread more than 200 km². The cleanup was mitigated by high seas causing the oil spill to break up at little cost and no damage to the environment (Hosseini, 2018).

| 7 | *Castillo de Bellver* | August 1983 | Cape Town, South Africa | Tanker catching fire | >78 | >1.8 |

Castillo de Bellver, carrying 252,000 tons of light crude oil, caught fire about 70 miles west of Cape Town, South Africa on August 6, 1983. The blazing ship drifted offshore and broke in two. The stern section – possibly with 100,000 tons of oil remaining in its tanks – sank in deep water. The bow section was towed away from the coast and was eventually sunk with the use of controlled explosive charges (ITOPF, 2020).

(Continued)

TABLE 4.2 (CONTINUED)
Summary of Oil Spills Greater than One Million Barrels since 1980

Order	Name	Date	Location	Cause	Approximate Amount of Oil Released	
					Million Gallons	Million Barrels (Mbbl)
8	*Amoco Cadiz*	March 1978	Portsall, France	Tanker runs aground	>68	>1.6

The tanker *Amoco Cadiz* ran aground off the coast of Brittany on March 16, 1978, following a steering gear failure. Over a period of two weeks the entire cargo of 223,000 tons of light crude oil and 4000 tons of bunker fuel was released into heavy seas. Much of the oil quickly formed a viscous water-in-oil emulsion, increasing the volume of pollutant by up to five times. By the end of April, oil and emulsion had contaminated 320 km of the Brittany coastline, and had extended as far east as the Channel Islands. Strong winds and heavy seas prevented effective offshore operation. About 3000 tons of dispersants were used (ITOPF, 2020).

Order	Name	Date	Location	Cause	Million Gallons	Million Barrels (Mbbl)
9	*Haven*	April 1991	Off the coast of Genoa, Italy	Explosion in the ship	~50	~1.2

This was the largest oil spill in the Mediterranean ever as a result of poor maintenance and an explosion in the ship which broke into three sections killing six crewmembers and contaminating Italian, French, and Monaco shores and coastline. The ship (VLCC) was loaded with 230,000 tons of crude oil of which about 144,000 tons were lost to the sea. The cleanup operations ended in 2008. The Italian government received total 120 billion Liras in compensation (Hosseini, 2018). More information is given in Section 4.4.2.

Order	Name	Date	Location	Cause	Million Gallons	Million Barrels (Mbbl)
10	*Odyssey* tanker	November 1988	Newfoundland, Canada	Storm breaking tanker	>43	>1

The accident occurred in the North Atlantic 700 miles off the coast of Nova Scotia, Canada. The ship sank as a result of rough weather conditions which was helpful in breaking oil particles so no damage to the environment was reported (ITOPF, 2020).

Order	Name	Date	Location	Cause	Million Gallons	Million Barrels (Mbbl)
11 (two spills)	*Atlantic Empress* and *Aegean Captain*	July 1979 and August 1979	Trinidad and Tobago/ Barbados	Two tankers collide/while being towed away	>42.7 and >41.5	~1 and ~1 (total 2)

(Continued)

TABLE 4.2 (CONTINUED)
Summary of Oil Spills Greater than One Million Barrels since 1980

					Approximate Amount of Oil Released	
Order	Name	Date	Location	Cause	Million Gallons	Million Barrels (Mbbl)
In July 1979, two VLCCs, the *Atlantic Express* and *Aegean Captain*, collided 10 miles off Tobago during a tropical storm. Both vessels caught fire and seven crewmembers died. Fire on the *Aegean* was brought under control but later a large explosion caused the *Atlantic Empress* to sink in August 1979 releasing further oil. An estimated 2 million barrels of crude oil spilled on these two occasions. This was the largest oil spill originating from an oil tanker (ITOPF, 2020).						
12	Production Well, D-103	August 1980	Tripoli, Libya	Well blowout	~42	~1
Although there is almost no detailed information regarding this spill, it is known that it occurred in August 1980, following a well blowout, with the release of approximately 140,000 tons of oil. The production well was located approximately 800 kilometers southeast of Tripoli, Libya. This was the largest oil spill originating from offshore installations in the Mediterranean Sea (EMSA, 2013).						
13	*Sanchi* tanker	January 6, 2018	260 km off Shanghai, China	Collision of two vessels	>42	>1
Collision of the *Sanchi* oil tanker with another vessel in January 2018 near Shanghai, China spread four separate oil spills covering 100 km^2. *Sanchi* was carrying Iranian condensate oil when it collided, killing 32 crewmembers onboard (BBC, 2018). This oil was a gas condensate and various sources give different amounts of oil from 113,000 to 150,000 tons equivalent to volume oil spill of greater than 1 million barrels. Further detail about this accident is given in Section 4.4.1.						

Source: ITOPF (2020).

90   Oil Spill Occurrence, Simulation, and Behavior

(the first five items) while spills caused by tankers constitute smaller volumes. However, the number of spills caused by tanker accidents is larger as was shown in Figure 4.1. The biggest oil spill by a tanker accident as shown in this table was the *SS Atlantic Empress* collision in the Mediterranean, causing two oil spills each about 1 million barrels. The total amount spilled was 287,000 tons in July and August, 1979. Other major spills caused by tankers include the *Castillo de Bellver, Amoco Cadiz, Haven, ABT Summer, Odyssey* tanker, and *Sanchi* tanker oil spills. One of the oil spills caused by a tanker carrying oil in South West Alaska when it ran aground was the *Exxon Valdez* oil spill which occurred in April 1989, and released about 37,000 tons of oil into the sea. Although the size of the spill was smaller than those given in Table 4.2, it had such an enormous impact on the environment and the oil industry that still after 30 years it is considered as one of the major oil spill accidents, and for this reason more information about this spill is given in Section 4.3.

In the next part (Section 4.4) three major oil spills caused by tankers introduced above are reviewed with more details on their specifications and impacts. In Section 4.5 the Persian Gulf oil spill and in Section 4.6 the Deepwater Horizon (BP) oil spill caused by the blowout of the Macondo well are discussed as the world's largest oil spills during last three decades.

4.4 TANKER OIL SPILLS

In this part three oil spills caused by tanker collision/allision are reviewed. These are the *Sanchi, Haven,* and *Exxon Valdez* tanker oil spills.

4.4.1 *SANCHI OIL SPILL*

The *Sanchi* oil spill is perhaps one the worst tanker spills in several decades. *Sanchi*, an Iranian oil tanker, collided with a Chinese Hong Kong cargo *Crystal* in the East China Sea 160 miles off the coast of Shanghai, China, on January 6, 2018, and sank on January 14 (Figure 4.4 and Figure 4.5).

The amount of oil released into the water reported in various sources ranges from 113,000 to 150,000 tons, but 136,000 tons is more frequently reported. It also killed 32 crew members of the tanker. In addition, 2,000 tons of bunker oil were also released which increased the total amount of the spill to 138,000 tons. Bad weather conditions with heavy rain and waves of up to 4 m hampered the rescue operations (BBC, 2018a). The API gravity of condensate oils is generally ≥ 45 or specific gravity ≤ 0.8. For this reason the conversion factor between metric tons and barrels is 1 ton $\approx$ 7.86. Using this factor, 136,000 tons of condensate is equivalent to about 1,070,000 barrels.

The tanker was carrying very light condensate and for this reason it was toxic and soluble in water. The slick area expanded to more than 330 km^2 after two weeks, although some models predicted an area of more than 350 km^2 after two months. However, most of the oil evaporated or formed an emulsion with water due to high waves. The tanker on fire just before it sank is also shown in Figure

Major Oil Spills

4.5 (ITOPF, 2018 and SCMP, 2018). It was believed that human error played a role in the accident as a result of lack of sufficient training of the pilot. The area in which the accident occurred is a busy area between China and Japan. For very light oils such as gas condensate the impact of the spill on air pollution is greater than water pollution as most of the oil evaporates quickly. Also the surrounding areas become much more combustible and flammable than with spills of heavy oil.

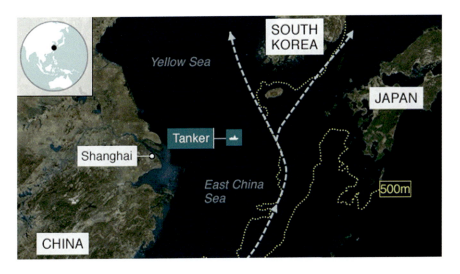

FIGURE 4.4 Map and location of Sanchi tanker collision (16 January 2018) (source: www.industryabout.com/industrial-news/).

 SANCHI INCIDENT

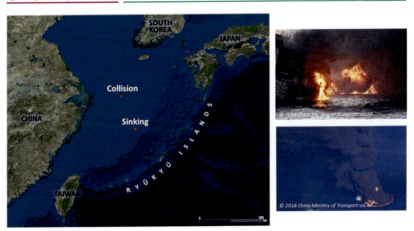

FIGURE 4.5 Sanchi tanker collision and sinking (ITOPF, 2018).

4.4.2 Haven Oil Spill

The *Haven* or MT *Haven* refers to the Cypriot motor tanker (MT) called *Haven* that on April 11, 1991, sustained a series of explosions and caught fire off the coast of Genoa, Italy. The cause of the explosions was apparently an electric spark during tank cleaning.

The ship was carrying 144,000 tons of heavy Iranian oil (1 million barrels) with API gravity of 31. The ship was a VLCC 334 m length, 54 m beam, and 26 m height with a capacity of 232,000 DWT (Amato, 2003). It is believed that about half of the oil was burnt during the fire (Figure 4.6). The spill contaminated the shores of Italy, France, and Monaco.

Analysis of the fate of the oil spill indicated that approximately 16,000 tons evaporated, 100,000 tons burnt, 1,000 tons washed ashore, 2000 tons were collected from the sea surface, 10,000 tons sunk, 3,000 tons in the wreck, and 4000 tons dispersed

FIGURE 4.6 MT *Haven* oil spill in the Mediterranean in April 1991. a. *Haven* tanker on fire (ITOPF, 1991). b. MT *Haven* tanker on fire (source: Amato, 2003).

at sea. Investigation of the sea bottom near the main wreck indicated that an area of about 120,000 m^2 was affected by a tar deposit 10 cm thick. At the sea surface, the concentration of polycyclic aromatic hydrocarbon (PAH) in water was about 0.366 μg per gram of water. These types of hydrocarbons are highly toxic. The fishing area was reduced, and as a result a 43% decrease in fish captures from the year before the incident was registered (Amato, 2003).

4.4.3 EXXON VALDEZ OIL SPILL

This accident occurred on March 24, 1989, when an American oil tanker *Exxon Valdez* ran aground at a speed of more than 22 km/h on Bligh Reef located 10 m deep waters of the Prince William Strait in Alaska, US. The accident damaged 11 tanks of 18 and spilled about 38,500 tons (equivalent to 11 million gallons or 262,000 barrels) of oil into the sea. The ship was 300 m in length and 50 m wide as shown in Figure 4.7. This was the biggest oil spill accident in US history until the 2010 Gulf of Mexico oil spill. The spill covered an area of more than 7000 km^2 and polluted about 2000 km of coastline (Figure 4.8) (Cedre, 2014).

The response to the oil spill involved the employment of 11,000 workers, and Exxon Mobil paid about $4.5 billion including cleanup costs and legal settlements (ITOPF, 2020; Cedre, 2014). Some 250,000 seabirds, 2,800 otters, 300 seals, 250 bald eagles, and about 22 whales were killed as a result of this accident. Shoreline cleanup techniques included high-pressure hot-water washing, which was carried out on a scale never attempted previously (Figure 4.9).

As reviewed by Riazi (2010) in response to the *Exxon Valdez* accident, the Oil Pollution Act (OPA) was signed into law in August 1990. The OPA improved the country's ability to prevent and respond to oil spills by expanding the federal government's ability and providing money and people if necessary. The OPA also created the national Oil Spill Fund which is available to provide a fund of up to 1 billion dollars per incident. In addition OPA increased penalties for regulatory noncompliance,

FIGURE 4.7 *Exxon Valdez* oil spill in Alaska right after the accident, 1989 (Cedre, 2014, original source: NOAA).

FIGURE 4.8 Three views of *Valdez* oil slick in 1989 (Cedre, 2014).

FIGURE 4.9 *Valdez* shoreline cleanup with high-pressure hot water (Cedre, 2014).

Major Oil Spills

and broadened the response and enforcement authorities of the federal government. In 1994, the EPA finalized a set of revisions to the OPA regulations that require facility owners or operators to prepare, and in some cases submit to the federal government, plans for responding to a major oil spill incident. For example, this law says the owners of oil tankers must have a detailed plan on what they will do if there is a spill. This law also says that all ships in the US are required to have a double hull by 2015. The law says the owners of a boat that spills oil will have to pay $1200 for every ton they spill. The law also allows the government to collect money from companies that transport oil so when a spill occurs the government can pay for the cleanup (EHSO, 2008).

4.5 PERSIAN GULF OIL SPILL

In Table 4.2 two oil spills are included from the Persian Gulf. The Nowruz oil spill occurred in 1983 as a result of Iraq–Iran war but the major oil spill occurred in January 1991 when UN coalition forces began expelling Iraqi forces out of Kuwait. Some oil fields in Kuwait were set on fire and oils from tankers were released into the sea (Figure 4.10). Initially it was estimated that some 11 million barrels were released into the Gulf but later the volume of oil spill was reduced to 5–8 million and in most references somewhere between 6 and 8 million was reported from government or private sources. Government researchers' estimates were always higher than private researchers' reported estimates. It was later reported that about half of the oil evaporated and about 2 to 3 million barrels washed ashore, mainly in Saudi Arabia. The slick reached a maximum size of about 100 miles in length and 42 miles in width with thickness in some areas about 5 in (13 cm). Certainly this spill is still considered as the largest oil spill in human history. Some pools of onshore spilled oil from the 1991 war still exist in parts of Saudi Arabia and Kuwait.

(a)

FIGURE 4.10 Persian Gulf Oil Spill, 1991. (a) Kuwait oil fields set on fire in January 1991 (Bechtel, 2020). (b) A Coast Guard Falcon jet crew monitors oil rigs set on fire in 1990 during the Gulf War (US Coast Guard photo). (c) Persian Gulf oil spill, January 23, 1991 (https://greasebook.com/blog/this-week-oil-gas-history-january-23/).

The Persian Gulf area is surrounded by many oil-producing countries with ports to export crude oil or its products. For this reason, it is an area with a very high risk of oil pollution in the sea and on the land. On August 12, 2017, an oil spill occurred in the northern Persian Gulf which affected the shores of Kuwait and Saudi Arabia (Figure 4.11). It was followed by several other spills with unknown amounts of oil

Major Oil Spills

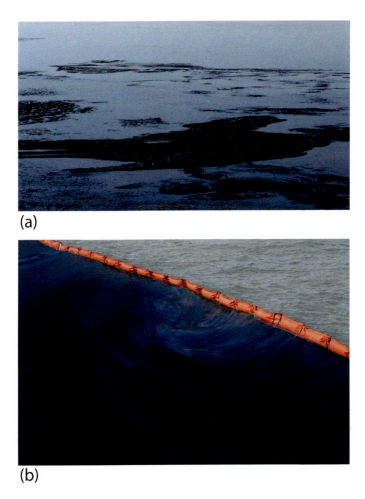

FIGURE 4.11 Kuwait – Persian Gulf Oil Spill in 2017. (a) Scene of an oil spill caused by a refinery on August 12, 2017 (www.meed.com, 2017). (b) Containment of Kuwait oil spill in Persian Gulf (PTJ, 2017).

released to the sea. The size of the first spill alone was 35,000 barrels of crude oil. The length of the second spill reached 1.6 km (PTJ, 2017). Authorities in Kuwait did not reveal the cause of the accident nor the exact amount of oil released to the sea, but later Chevron and local agencies became involved in the cleanup operation and successfully contained the spill (Sen, 2017). This was the most recent major oil spill in the region with limited data available in the open literature about the nature of spill, its causes, the amount, and its fate in the sea. A summarized version of major oil spills excluding the 1991 Persian Gulf spill is shown in Figure 4.12 (BBC, 2010).

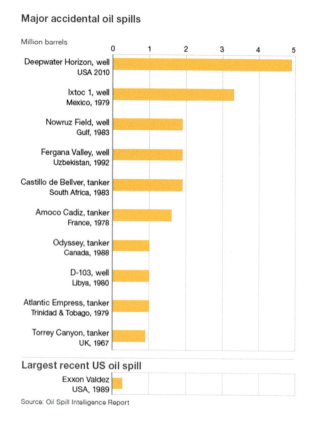

FIGURE 4.12 Largest oil spills as reported by Richard Black of BBC on 3 August 2010.

4.6 DEEPWATER HORIZON (GULF OF MEXICO) OIL SPILL 2010

On April 20, 2010, the Deepwater Horizon drilling rig exploded in the Gulf of Mexico, killing 11 workers and becoming the worst environmental disaster in US history. The drilling rig was operated by Transocean while the exploratory Macondo well was mainly owned and operated by BP. For this reason the oil spill is referred to as the Deepwater oil spill, Gulf of Mexico (GOM) 2010 oil spill, or BP oil spill. In this book generally the accident is referred to as the BP oil spill and the dates cited without year refer to 2010.

As this is the most recent and major environmental disaster that occurred in the US and it was hugely covered by the media during 2010 and the years that followed, information about this accident was available more than any other similar events in the past. As this book is being prepared during the tenth anniversary of that disaster, the environmental, economic, and the political impacts as well as the cleanup methods are reviewed in this chapter and continue in Chapters 5 and 6.

Major Oil Spills 99

4.6.1 Chronology of Events

The timeline of events and breaking news from April 20 until September 19, 2010, as reported by various news agencies is presented here. The story and the data given vary depending on the reporting source such as US government, BP, US media, or British media. Here is a chronology of events as it was reported by BBC World on September 19, 2010.

20 April. An explosion happened on the Deepwater Horizon drilling rig in the Gulf of Mexico about 50 miles south of Venice, Louisiana. The rig was drilling about 5,000 ft below the water surface never experienced before. Eleven workers on board were killed and 17 were injured.

22 April. The Deepwater Horizon sinks to the bottom of the Gulf after burning for 36 hours. A Coast Guard official said the Macondo well could be releasing some 8,000 b/d of oil.

23 April. The officials could not confirm how much oil was coming to the water surface.

26 April. The first official estimate of the amount of oil released is 1000 b/d, and they warn of a major environmental disaster. Meanwhile BP try unsuccessfully to activate the blowout preventer (BOP) to stop the oil flow.

28 April. The US Coast Guard warn the leak could be the worst oil spill in US history.

29 April. The US Coast Guard sets fire to the spilled oil to prevent it reaching the coastal wetlands of Louisiana. President Obama pledges every possible resource including use of the military to control the spill.

30 April. Oil begins washing ashore in Louisiana. The US administration bans oil drilling in new areas off the US coast pending investigations on the cause of the BP oil spill.

2 May. President Obama makes his first trip to the Gulf Coast and says BP is responsible for the leak and for paying its cleanup operations.

8 May. BP tries unsuccessfully to contain the spill by placing a giant metal box on the top of the leaking well due to the formation of ice crystals inside the box. Officials revise the oil rate to 5000 b/d.

10 May. BP officials weigh inserting debris and rubber tires in the wellhead, a method known as the "junk shot." BP also announce the cost of spill over $350 million so far.

11 May. At a series of congressional hearings, BP, Transocean, and Halliburton, the three major companies involved in the Deepwater Horizon drilling operations, all blame each other for the disaster.

14 May. Researchers who have analyzed underwater video from the leak site estimate that as many as 70,000 barrels of oil are leaking into the Gulf per day, with a margin of error of plus or minus 20%, significantly higher than earlier estimates.

19 May. Oceanographers say oil from the leak has entered an ocean current – the "loop current" – that could carry the oil towards Florida.

26 May. BP prepares to plug the leaking well with heavy drilling mud, a procedure called a "top kill." The attempt is declared a failure three days later.

28 May. Obama visits the Gulf Coast again for the second time.

30 May. The president's energy adviser says the spill is the worst environmental disaster in US history, worse even than the 1989 *Exxon Valdez* spill in Alaska.

2 June. The US announces a criminal inquiry into the BP oil spill.

4 June. BP places a cap atop the leaking wellhead to allow much of the oil and gas to be piped to the ships on the surface. President Obama visits the region for the third time.

8 June. The US government says underwater oil plumes have traveled as far as 40 miles from the site of the leaking well. Officials also warn that the cleanup operations could take years.

10 June. The US Geological Survey estimates the oil flow at as high as 40,000 b/d before a cap was put on the well on 3 June. BP announces it is collecting 15,800 b/d from the well through the cap.

12 June. President Barack Obama and UK Prime Minister David Cameron discuss possible impacts of the oil spill on bilateral relations as reported in the media.

14 June. President Obama makes a fourth trip to the gulf.

15 June. President Obama addresses the nation from the Oval Office, saying to the American people, "We will make BP pay for the damage it has caused."

17 June. BP announces it will place $20 billion in a fund to compensate victims of the oil spill and says it will not pay dividends to shareholders in 2010.

18 June. BP CEO Tony Hayward receives a tongue-lashing at a hearing in the US Congress.

22 June. BP hands day-to-day control of the response to Bob Dudley, replacing Chief Executive Tony Hayward, who finally takes the job as the chief executive.

5 July. BP says the oil spill response has cost the company over $3 billion so far.

6 July. Oil from the spill reaches Texas, meaning it has affected all five US Gulf Coast states.

10 July. BP begins a bid to place a tighter fitting cap atop the leaking wellhead. The company warns that oil flow will increase for a few days, but says it has brought in 400 oil-skimming ships to deal with the increased oil flow.

14 July. BP and US officials say a relief well has come within 5 ft of the leaking well and say this is the only way to plug the well permanently.

15 July. With the new cap in place, BP says it has temporarily shut off the oil flow. President Barack Obama hails it as a "a positive sign."

19 July. A US official is concerned about a "detected seep" on the sea floor near the well.

Major Oil Spills

22 July. Due to an expected tropical storm some activities are delayed. BP says it has been given permission to prepare a "static kill" while waiting for the final approval of the plan from US officials.

25 July. Ships involved in BP's effort to secure the blown-out oil well prepare to resume work after a tropical storm in the Gulf of Mexico weakens.

26 July. The BBC reveals that 53-year-old BP Chief Executive Tony Hayward will receive a year's salary plus benefits, together worth more than £1 million, when he steps down.

27 July. BP confirms that Tony Hayward will leave his post as chief executive by October. BP announces a loss of $17 billion for the second quarter of the year.

28 July. Some US scientists say the oil from the well has cleared from the sea surface faster than expected.

2 August. The US Environmental Protection Agency says in a study that the dispersant used after the spill is no more toxic than oil alone.

3 August. The US government says the oil spill is officially the biggest leak ever, with 4.9 million barrels of oil leaked before the well was capped last month. Scientists say only a fifth of the leaking oil – around 800,000 barrels – was captured during the cleanup operation.

4 August. The US government says three-quarters of the oil spilled has been cleaned up or broken down by natural forces. BP reports good progress with the static kill to plug the well with mud and seal it with cement permanently.

9 August. BP announces that the total cost to it of the oil spill so far has reached $6.1 billion including the cost of the spill response, containment, relief well drilling, and cementing up of the damaged well.

16 August. The US announces that future applications for deep-water offshore drilling will require an environmental assessment.

19 August. A study published in a leading scientific journal confirms the presence of a toxic chemical residue 1 km below the surface of the Gulf of Mexico.

3 September. The blowout preventer (BOP) that failed to stop the explosion is removed from the oil well by BP to be examined for the cause of failure. BP announces the cost of the oil spill raised to $8 billion.

5 September. The US official overseeing the cleanup operation says the oil well no longer poses further risk to the environment.

8 September. BP in its own internal report accepts responsibility in part for the disaster.

17 September. BP pumps cement to seal the damaged well after it was intercepted by a relief well.

19 September. The ruptured well is finally sealed and "effectively dead," says a top US federal official overseeing the disaster.

4.6.2 How it Occurred

The Deepwater Horizon drilling rig (Figure 4.13) was owned by Transocean and operated and contracted to BP on the Macondo Prospect when an explosion occurred at 10 pm on Tuesday April 20, 2010, in the Gulf of Mexico (Figures 4.14, 4.15). The top of the well was about 5,000 ft beneath the surface of the Gulf of Mexico.

The Macondo well is located about 50 miles southeast of Venice, Louisiana (Figure 4.16). **BP** was the operator and principal developer of the Macondo Prospect with a 65% share, while 25% was owned by Anadarko Petroleum, and 10% by MOEX Offshore (Guardian, April 21).

At the time of the explosion 126 people (the majority from Transocean) were on board where 11 of them were killed and 17 people were injured. The surface facilities were surrounded by gases entering the engine rooms of Deepwater, causing an explosion. The gases came from an uncontrolled upward surge of oil and gas to the surface as a result of the blowout preventer (BOP) failure. The BOP, the size of a five-story building, consists of a series of high-pressure valves designed to prevent

FIGURE 4.13 Deepwater Horizon before the accident, located 51 miles southeast of Venice, Louisiana (Deep Water, 2011).

Major Oil Spills

FIGURE 4.14 (a) The Deepwater Horizon burns on April 21, 2010, with fire boat response. The drilling rig sank the next day. (U.S. Coast Guard photo: © USCG). Fire boat response crews battle the blazing remnants of the off shore oil rig, Deepwater Horizon on April 21, 2010. The rig, located 51 miles southeast of Venice, Louisiana, exploded on April 20, 2010. (Image: © USCG). (b) Arial view of the rig on April 21, 2010. (Courtesy of AP). Deepwater Horizon rig after the explosion.

FIGURE 4.15 Gulf of Mexico, May 16, 2010 (photo by Petty Officer 2nd Class Patrick Kelley US Coast Guard Atlantic Area [USCG-AA]).

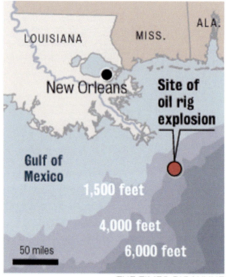

FIGURE 4.16 Location of explosion.

Major Oil Spills 105

FIGURE 4.17 The rig's blowout preventer failed to prevent a massive oil leak (Courtesy of BP).

such a gas surge as shown in Figure 4.17 (BBC). Investigations into the reason for the BOP failure were focused on the valves as shown in Figure 4.18. The failed BOP was brought to the water surface on September 4, 2010, as shown in Figure 4.19. Later survivors told CNN that on the morning of the accident there was an argument over how to proceed with the drilling between BP executives and Transocean officials. Rescue boats began spraying on the rig the next day (Figure 4.20).

BP said the oil flowing from well was brought to a ship while the company was trying to close four valves of the BOP to stop the flow as shown in Figure 4.21 (CNN, June 4). On June 8 the government officials put an estimate of oil flow rate up to 43,000 b/d (1.8 million gallons/d).

An oil slick off the Mississippi Delta due to the Deepwater Horizon disaster is shown in Figure 4.22a. Figures 4.22b shows a satellite image of the spill on April 25 taken by NASA's Aqua and using its Moderate Resolution Imaging Spectroradiometer

106 Oil Spill Occurrence, Simulation, and Behavior

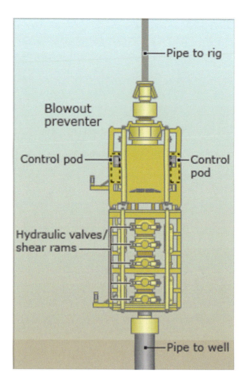

FIGURE 4.18 Investigations into the US oil spill were focused on the blowout preventer system of valves on the seabed (BP graphic, BBC September 8).

FIGURE 4.19 A Coast Guard photo shows the failed blowout preventer being raised from the Macondo well site on September 4 (USCG/BBC).

Major Oil Spills 107

FIGURE 4.20 Rescue boats spray the fire on the rig on April 22 (image by US Coast Guard, https://blog.response.restoration.noaa.gov/).

FIGURE 4.21 Oil is flowing from damaged well in GOM (www.whoi.edu).

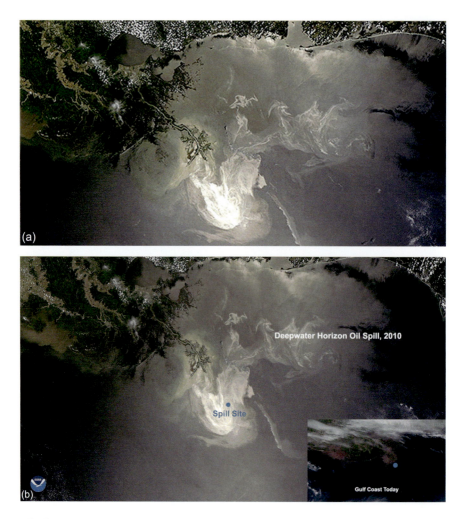

FIGURE 4.22 (a) An oil slick off the Mississippi Delta on May 24, 2010, due to the Deepwater Horizon disaster (source: NASA). (b) Satellite image of the Gulf Coast on April 25 (NOAA).

(MODIS) instrument, showing the Mississippi Delta on the left and a wide oil spill on the right. Figure 4.23 shows an aerial view of weathered oil near the Louisiana shore, and Figure 4.24 shows an infrared image of the slick taken by NASA in May 2010. Figure 4.25 shows a plume of oil leaking from the well, and Figures 4.26 and 4.27 show aerial images of the slick in the early days after the accident. A series of

Major Oil Spills

FIGURE 4.23 An aerial image of Gulf of Mexico oil slick, May 17, 2010 (source: UPI/Marc).

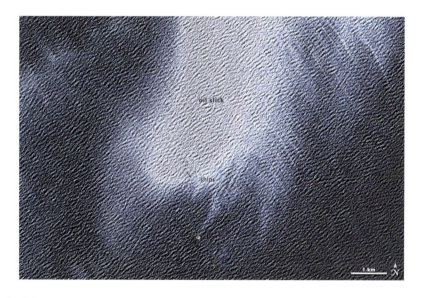

FIGURE 4.24 Infrared image of the BP oil slick in May 2010 (source: NASA/Flickr).

FIGURE 4.25 A plume of oil leaking into the Gulf of Mexico at British Petroleum's Deepwater Horizon oil spill site, July 13, 2010 (source: UPI/BP).

FIGURE 4.26 An aerial image of the oil spill in early days of the accident (NASA photo).

images taken between April 21 and May 24, 2010, is presented in Figure 4.28, and it shows the growth of the oil slick from the Deepwater Horizon oil rig explosion in the Gulf of Mexico (UPI/NASA/MODIS Rapid Response).

Figures 4.29, 4.30, and 4.31 show the trajectories of oil movement based on overflight information in the Gulf of Mexico from April through August 2010 as simulated by NOAA. Another simulation of the oil spill by NOAA in May 2010 is shown in Figure 4.32.

Major Oil Spills 111

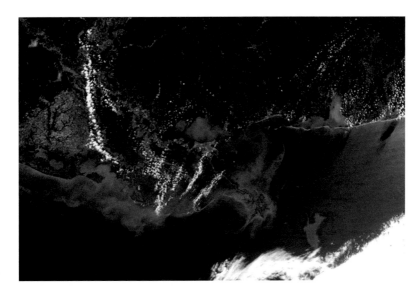

FIGURE 4.27 Satellite image of the Gulf of Mexico oil spill on May 4, 2010 (source: NOAA).

Series of NASA images showing growth of BP oil spill

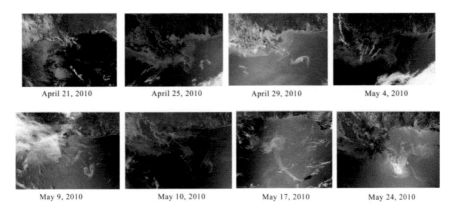

FIGURE 4.28 A series of NASA satellite images showing growth of the BP oil spill (NASA).

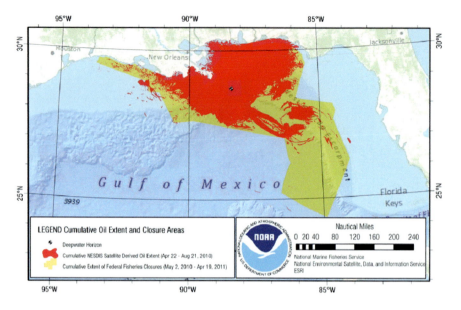

FIGURE 4.29 Map of Gulf of Mexico showing the extent of the oil spill that occurred from April 22 through August 21, 2010. The black-and-white circle indicates the wellhead location. The orange-shaded areas show the cumulative National Environmental Satellite, Data and Information Service (NESDIS) satellite footprints of the oil. The yellow polygon overlay shows how the federal fisheries closure areas aligned with the oil distribution (www.researchgate.net).

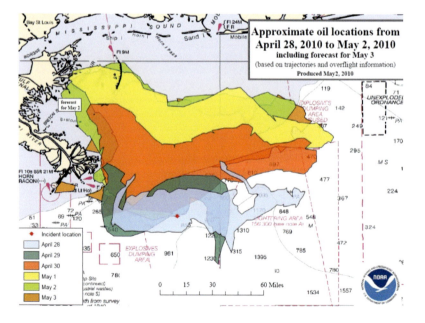

FIGURE 4.30 Approximate oil locations from April 28, 2010 to May 2, 2010 (source: response.restoration.noaa.gov/).

Major Oil Spills

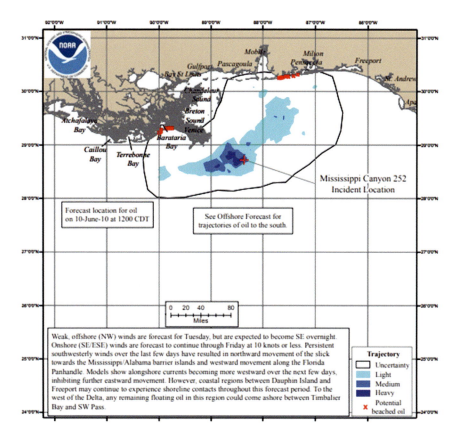

FIGURE 4.31 BP oil spill trajectory (NOAA, June 7, http://response.restoration.noaa.gov).

Figure 4.33 shows how the oil spill approaches the coast of Mobile, Alabama on May 6, 2010, as photographed by the US Navy. An oil sheen from the BP oil spill in May 2010 as photographed by the US Department of Interior (DOI) is seen in Figure 4.34.

As will be discussed in Chapter 5, the BP oil spill caused significant damage to the environment as some fish were seen near the 21-in erupted pipe 5,000 ft below the water surface. BP executed its "Top Kill" process, which places heavy kill mud into the oil well in order to reduce pressure and the flow of oil from the well as shown in Figure 4.35 which is a grab from the video of the operation. The method was not successful simply because the pressure of oil and gas from the well was too powerful to overcome. In Section 4.6.7 it is demonstrated how the flow of oil was eventually stopped by mid-July and the well was permanently dead in September.

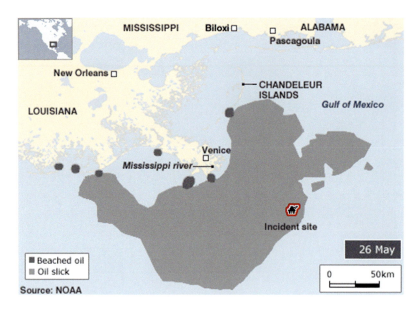

FIGURE 4.32 Trajectory of spill as simulated by NOAA on May 26 (NOAA).

FIGURE 4.33 Oil from the Deepwater Horizon oil spill approaches the coast of Mobile, Ala., May 6, 2010 (US Navy/Wikimedia.org).

Major Oil Spills

FIGURE 4.34 DOI photo of oil sheen from BP oil spill on May 5, 2010 (www.doi.gov).

FIGURE 4.35 The unsuccessful procedure known as "Top Kill" began May 26, 2010 (UPI/BP).

4.6.3 Deepwater Horizon, Macondo Well, and Fluid Composition

In 2011 the total oil production from Gulf of Mexico was about 23% of US oil production while in 2019 this number was 15%, as onshore production growth continues to outpace offshore production. However, the rate of oil production from the GOM is at a record high at 2 million barrels per day (EIA, 2019).

In 1901 oil was nearly at the surface of land in the state of Texas, and when land-based oil was exhausted oil producers went to the sea. In 1985 only 21 million

FIGURE 4.36 Gulf of Mexico, USA – July 31, 2009: TransOcean's oil rig; before the accident (source: Istockphoto.com).

TABLE 4.3
Composition of Gas, Oil, and Reservoir Fluid from Oil of MW-1 from the Macondo Well on June 21, 2010

Compound	Gas Released, Mol%	Oil Released, Mol%	Reservoir Fluid Mol%
CO_2	1.19	0	0.98
C_1	80.96	0	68.48
C_2	8.33	0	6.80
C_3	5.21	0	4.31
iC_4	0.93	0.38	0.84
nC_4	1.87	1.82	1.89
iC_5	0.52	2.52	0.80
nC_5	0.49	4.09	1.01
C_6	0.50	18.12	3.20
C_{7+}	0	73.06	11.70

barrels or 6% of oil produced in the GOM came from wells in water more than 1,000 feet deep. In 2009 these wells produced 456 million barrels or 80% of the Gulf production (Newsweek, June 14, 2010).

The DWH rig for drilling the Macondo well (an exploratory well) before the accident is shown in Figure 4.36 and was owned by Transocean operated by BP. The site was located 52 miles southeast of Venice, Louisiana, the southernmost town on the Mississippi River. The underground formation that the rig was probing (owned by BP) has been estimated to hold anywhere from 50 million to a billion barrels of oil that could be extracted economically (ABC June 4).

The specifications of reservoir and the fluid are given below:

The reservoir formation pressure was 13,000 psi, and the pressure below BOP was at 8,000–,9000 psi, formation temperature at 180–200°F, reservoir fluid density in the range of 700 to 860 kg/m³, water pressure at the bottom of sea (hydrostatic pressure)=2300 psi, injected mud density for static kill=12–14 lb/gallon versus water density of 8 lb/gallon, temperature at the bottom of the sea=30°F.

The composition of fluid is given below.

The most readily available data for the crude oil produced from the Macondo well are its density or API gravity. However, Reddy et al. (2011) reported a detailed GC analysis of oil and gas produced from the well at 5000 feet below water surface based on a sample taken on June 21, 2010. They found that the fluid flowing out of the Macondo well had a gas-to-oil ratio (GOR) of 1,600 standard cubic feet per barrel. The composition of the oil, gas, and reservoir fluid is given in Table 4.3 while general characteristics of the oil are given in Table 4.4. The oil API gravity is 40 (equivalent to density of 0.85 g/cm³) as reported by Reddy (2011) and US government organizations (NOAA, 2010). Based on the normalization of data given in Table 4.3, the molar distribution of this crude oil was generated which is used for the simulation of the rate of vaporization and dissolution in Chapter 7. These data are calculated from measurements reported by Reddy et al. (2011)

TABLE 4.4

Oil General Characteristics

API gravity	40
Carbon %	86.6
Hydrogen %	12.6
Nitrogen %	0.38
Sulfur %	0.39
Saturated hydrocarbons %	74
Aromatic hydrocarbons %	16
Polar hydrocarbons %	10

4.6.4 METEOROLOGICAL DATA

As shown in Chapter 6, data on wind speed and air temperature are needed in order to determine the rate of vaporization (Riazi, 2012). These data changed on a daily or hourly basis from the time of the accident until the well was sealed on July 15, 2010. Full meteorological data were obtained from a Weather Channel website (www .weather.com) for Venice, Louisiana (Weather, 2010), for the period in which oil was flowing into the sea. Table 4.5 gives a sample of data on temperature and wind speed variation in 24 hours on May 28, 2010. Daily data are given in Table 4.6 until August 9. Days are counted from April 20, 2010, for 111 days.

Low and high temperatures on a daily basis are also shown in Figure 7.15 and presented by a quadratic equation in Table 7.5 (see Chapter 7). Humidity data from Table 4.6 are also presented in Figure 4.37 (Riazi, 2012, 2016). These data will be used in the calculations presented in Chapter 7.

TABLE 4.5

Variation of Temperature, Humidity, Wind Speed, and Cloud Overage in Venice, LA During May 28, 2010

Time, hr	Temp, °C	Dew Point, °C	Humidity %	Wind, km/h	Wind Direction Direction	Wind Direction Degrees	% Cloud Coverage	% Chance of Precipitation
0	26	21	76	13	WSW	250	45	10
4	26	21	74	13	WSW	280	44	10
7	26	21	74	13	WSW	280	43	10
10	29	22	64	10	NNW	330	43	20
13	29	22	65	3	WNW	290	52	20
16	30	22	63	6	SW	230	52	20
19	28	22	69	13	SW	230	38	20
22	26	22	79	10	WSW	240	38	20

Source: Weather Channel, www.weather.com.

Major Oil Spills

TABLE 4.6
Variation of Temperature, Humidity, and Wind Speed in Venice, LA During June–August 2010 Period Recorded at 1 am Local Time

Day from April 20	Day	Date	Wind, km/h	Wind Direction, From	Temperature, °C		Humidity %	Sky Condition*
					Low	High		
16	Thu	6-May	10	S	20	29	90	S
17	Fri	7-May	19	SW	21	27	74	S
18	Sat	8-May	19	SW	22	29	88	C
19	Sun	9-May	16	NE	19	25	57	C
20	Mon	10-May	8	E	21	27	87	S
21	Tu	11-May	18	SE	23	29	91	S
22	Wed	12-May	13	SE	23	30	93	S
23	Thu	13-May	21	SE	24	31	90	S
24	Fri	14-May	19	E	23	30	89	S
25	Sat	15-May	10	SE	23	29	90	PC
26	Sun	16-May	14	SE	23	28	78	C
27	Mon	17-May	6	S	23	29	95	S
28	Tue	18-May	14	SW	22	31	92	S
29	Wed	19-May	16	SW	23	29	80	S
30	Thu	20-May	19	S	24	29	67	S
31	Fri	21-May	2	S	24	30	88	PC
32	Sat	22-May	21	S	24	31	90	S
33	Sun	23-May	0	N	25	31	89	PC
34	Mon	24-May	0	N	24	32	85	S
35	Tue	25-May	0	N	24	31	88	S
36	Wed	26-May	8	E	23	29	76	PC
37	Thu	27-May	0	N	23	32	81	PC
38	Fri	28-May	14	W	24	33	78	S
39	Sat	29-May	10	W	25	33	74	PC
40	Sun	30-May	8	SE	24	31	86	PC
41	Mon	31-May	8	S	24	30	78	PC
42	Tue	1-Jun	8	S	24	29	83	PC
43	Wed	2-Jun	18	S	24	29	85	PC
44	Thu	3-Jun	11	SE	25	28	90	S
45	Fri	4-Jun	11	SW	26	29	82	PC
46	Sat	5-Jun	18	S	26	32	89	PC
47	Sun	6-Jun	6	W	26	33	87	C
48	Mon	7-Jun	6	SW	26	32	91	S
49	Tue	8-Jun	0	N	26	32	93	S
50	Wed	9-Jun	8	S	26	32	87	S
51	Thu	10-Jun	6	SE	27	30	87	PC
52	Fri	11-Jun	11	SE	27	32	88	S

(Continued)

Oil Spill Occurrence, Simulation, and Behavior

TABLE 4.6 (CONTINUED)
Variation of Temperature, Humidity, and Wind Speed in Venice, LA During June–August 2010 Period Recorded at 1 am Local Time

Day from April 20	Day	Date	Wind, km/h	Wind Direction, From	Temperature, °C Low	Temperature, °C High	Humidity %	Sky Condition*
53	Sat	12-Jun	14	SE	26	33	72	PC
54	Sun	13-Jun	14	SW	26	32	66	C
55	Mon	14-Jun	5	SW	27	33	81	PC
56	Tue	15-Jun	11	S	26	32	76	S
57	Wed	16-Jun	0	N	26	32	86	PC
58	Thu	17-Jun	2	W	27	33	94	S
59	Fri	18-Jun	6	SW	27	32	81	S
60	Sat	19-Jun	10	SW	27	32	67	S
61	Sun	20-Jun			27	33		
62	Mon	21-Jun			27	32		
63	Tue	22-Jun			27	31		
64	Wed	23-Jun	8	SE	27	31	90	PC
65	Thu	24-Jun			27	32		
66	Fri	25-Jun	0	S	27	31	82	PC
67	Sat	26-Jun	11	NE	27	33	82	S
68	Sun	27-Jun	5	E	27	32	85	S
69	Mon	28-Jun			27	32		
70	Tue	29-Jun			27	31		
71	Wed	30-Jun			27	30		
72	Thu	1-Jul	16	E	27	29	91	C
73	Fri	2-Jul			26	30		
74	Sat	3-Jul			26	30		
75	Sun	4-Jul			26	30		
76	Mon	5-Jul			26	30		
77	Tue	6-Jul	13	E	27	29	87	PC
78	Wed	7-Jul	31	SE	27	31	76	C
79	Thu	8-Jul			26	31		
80	Fri	9-Jul			27	32		
81	Sat	10-Jul			27	32		
91	Tue	20-Jul	0	N	28	31	88	S
92	Wed	21-Jul			28	31		
93	Thu	22-Jul			28	32		
94	Fri	23-Jul	13	E	27	32	62	S
95	Sat	24-Jul	29	SE	27	31	89	C
96	Sun	25-Jul	16	SE	27	34	80	S
97	Mon	26-Jul			26	31		
98	Tue	27-Jul			27	31		
99	Wed	28-Jul	11	SE	26	34	61	S

(Continued)

TABLE 4.6 (CONTINUED)
Variation of Temperature, Humidity, and Wind Speed in Venice, LA During June–August 2010 Period Recorded at 1 am Local Time

Day from April 20	Day	Date	Wind, km/h	Wind Direction, From	Temperature, °C Low	Temperature, °C High	Humidity %	Sky Condition*
100	Thu	29-Jul			27	33		
101	Fri	30-Jul			28	33		
104	Mon	2-Aug	5	NW	28	33	70	S
105	Tue	3-Aug			27	33		
106	Wed	4-Aug	8	NE	27	33	75	PC
107	Thu	5-Aug			27	33		
108	Fri	6-Aug			27	33		
109	Sat	7-Aug	13	N	28	32	75	C
110	Sun	8-Aug			27	33		
111	Mon	9-Aug			28	33		

Source: Weather Underground, http://www.wunderground.com/US/LA/Venice.html.
* Key for sky conditions (S: sunny, C: cloudy, PC: partly cloudy).

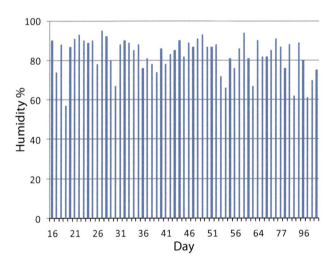

FIGURE 4.37 Variation of air humidity versus days (April 20, 2010, Day = 0).

4.6.5 Oil Flow Rate and Slick Area

4.6.5.1 Oil Flow Rate

Although throughout the disaster, from April to July, data on the oil flow rate and area covered by the oil were reported, usually they were inconsistent and revised with time. Generally three main sources were reporting the rate of oil released into the sea: BP, the US government, and independent scientists from research centers. Generally the rates reported by BP were lower than US government data and scientists reported higher rates, although toward the end of July the data got closer to each other and to those reported by the scientists. At the beginning the US government was mainly relying on the data provided by BP but later its data were mainly from the scientists based on underground images and simulation as they were more acceptable.

During the early days right after the accident BP announced the rate of oil flow to the Gulf was at 1000 b/d; however, after a week the number was raised to 5,000 barrels per day. As late as May 15, BP was insisting the flow rate was 5000 b/d, well below the actual rate, while the scientists on May 1 gave an estimate of 25,000 bbl/day. On May 27, 2010, the government increased its official estimate of 12,000–19,000 barrels per day.

For the first time scientists from University of Georgia announced that there was a shocking amount of oil under the sea. They said there were three to five layers of oil in the water column. After studying footage of the gushing oil, scientists on board the research vessel *Pelican*, which was gathering samples and information about the spill, said that the oil could be flowing at a rate of 25,000 to 80,000 barrels per day. The government estimate based on satellite images and BP data was at 5000 b/d. The scientists found that a plume of oil 10 miles long, three miles wide and 300 ft thick ($16 \times 5 \times 0.1$ km^3) was pouring into deep waters of the GOM from the ruptured oil rig (New York Times, May 16, 2010). In June, CNN reported that scientists from the College of Marine Science at the University of South Florida said they found evidence of another plume lurking underwater (CNN, June 7, 2010). BBC Radio said this plume could be 20–30 km long (May 17 at 5 am GMT). On June 11 CNN reported a new estimate of up to 40,000 b/d based on new images of gushing oil. The final rate that was given by the US government based on the scientists' estimate was 57,000 barrels per day (377 m^3/h). For the period of 87 days from April 20 to July 15, when the well was sealed, the total amount of oil released was estimated at 4.9 million barrels (NOAA). This estimate was also accepted by BP and reported by different sources as the net amount of oil gushed into the water. Generally various other sources put the rate at $5.6 \times 10^4 \pm 21\%$ bbl/day.

4.6.5.2 Surface Area

The surface area not only depends on the amount of oil floating on the water surface but also depends on the weather conditions, water current, wind speed, and properties of the oil. The most important property that determines the degree of spreadability of oil is its interfacial tension with water or simply the oil's surface tension. When the interfacial tension between oil and water is low, the oil can spread on the

Major Oil Spills

water surface easily and cover a higher area. As discussed in Chapter 2, light oils have lower surface tension, and for this reason they can cover more surface area than heavier oil under the same conditions. As temperature increases surface tension decreases, and for this reason in summer and under hot conditions the area covered by the oil is greater than what it could cover during the winter time when the temperature is lower.

BP, based on its own data, reported a slick area of 5200 km^2 on May 16. However 10 days earlier on May 7, NOAA reported that the spill covered an area of 4,000 square miles or 10,360 km^2, twice as much as the BP estimate on May 16. However, on June 7, BP confirmed that the area covered by the spill was about 10,000 km^2. Some US officials gave an area between 10,000 and 16,000 km^2 as of May 8.

Usually information about the spill surface area and sea conditions is obtained from satellite data. Figure 4.38 shows sea surface velocity from satellite radar data obtained on May 27 by the Colorado Center for Astrodynamics Research (CCAR) at the University of Colorado in Boulder. As shown in this image, the SSV at this location varied from 0 to 2 knots (equivalent to 3.7 km/h or 2.3 mph).

Figure 4.39 shows how the slick is spreading by a satellite image taken on May 22. According to this image the oil slick and sheen cover an area of 42,833 km^2. Advanced synthetic-aperture radar (ASAR) was used by the Center for Southeastern Tropical Advanced Remote Sensing (CSTARS) at the University of Miami for an image produced on June 3 as shown in Figure 4.40. Based on these data the surface

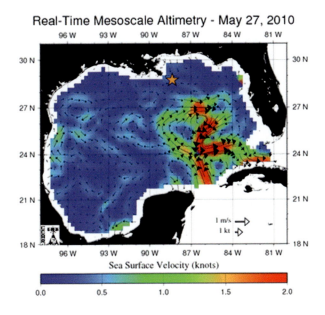

FIGURE 4.38 Sea-surface velocity (SSV) map derived from satellite radar altimeter data, May 27, 2010. Location of the loop current is indicated by green to red band of relatively high velocity at the ocean surface (source: https://skytruth.org/2010/05/bp-gulf-oil-spill-moving-toward-florida/).

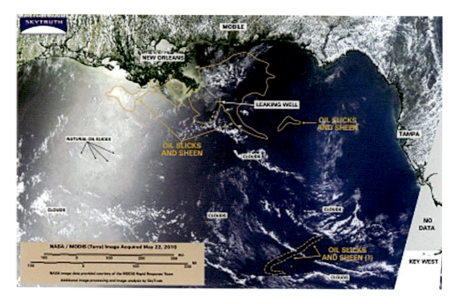

FIGURE 4.39 MODIS/Terra satellite image (Skytruth, May 22).

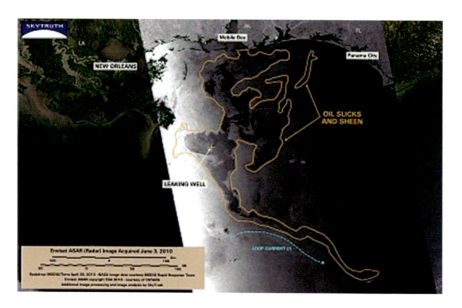

FIGURE 4.40 Envisat ASAR satellite radar image, June 3, 2010 (image courtesy CSTARS).

Major Oil Spills

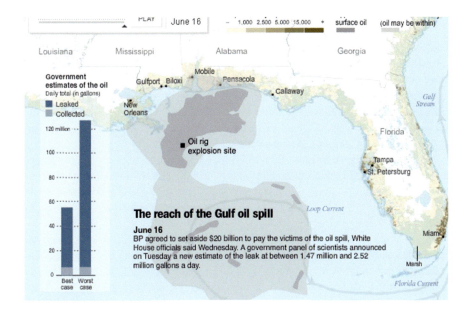

FIGURE 4.41 Oil track on June 16, 2010 (NOAA/Yahoo).

area of the slick was calculated at 29,796 km². The oil track as simulated by NOAA on June 16 is shown in Figure 4.41. In this figure, the best and worst estimates by the government of the amount of oil released are also shown.

Figure 4.45 (in Section 4.6.6) on June 7 (day 49) was used to calculate the areas as: 9220 for light, 875 for medium, and 54 km² for heavy oils. This gives a total estimated area of 10,150 km² which is very close to the reported area in early June by US government officials. This indicates that only 9.1% of total area is covered by medium and heavy oils. Assuming a flow rate of 28,000 b/d as given by BP on June 7, the slick thickness was estimated at 0.0215 mm. The actual oil flow rate was certainly higher than 28,000, but considering that some were burning and some were removed from the sea this number could be considered as an approximate amount of oil released to the water surface.

Based on the distance traveled by the spill and time an approximate velocity of oil movement can be obtained. For example, BP reported that after 7 days the spill had moved a distance of 50 miles (80 km). This gives that the oil was moving at a speed of 0.13 m/s which is much less than the average oil current of 0.5 m/s in that area. Similarly, based on the amount of oil floating on the seawater surface and the area covered by the slick, an approximate thickness of slick can be calculated.

The ruptured pipe had a diameter of 21 in or cross-sectional area of 0.2233 m². Based on the oil flow rate of 40,000 b/d or 0.0736 m³/s, this gives an upward velocity of 0.33 m/s for the oil to travel upward in the water column. Assuming oil density of 0.65 g/cm³ and viscosity of 0.5 mPa·s, the Reynolds number can be calculated as 2.3×10^5 which indicates the oil flow regime was turbulent.

4.6.6 Trajectory and Simulation

The National Oceanic and Atmospheric Administration (NOAA) is an American scientific agency within the Department of Commerce that focuses on the prediction of conditions of the oceans and was heavily involved in forecasting BP oil spill movements and behavior. Its daily forecast of the spill movement was reported by news agencies, especially Associated Press (AP), from April to July 2010 and was released by NBC, CNN, CBS, ABC, and BBC World news organizations. In addition, the National Center for Atmospheric Research (NCAR) was also involved in the modeling and forecasting of water streamflow. In addition several universities such as Texas A&M and Florida State University have developed simulators that predict the movement and behavior of oil spills on the water surface. Some of these models are presented in Chapter 8. In this section some predicted extend of oil slick as produced by NOAA and other centers are presented.

Figure 4.42 shows oil locations on May 6 (from overflight information) and May 7 as forecasted by NOAA (BBC). Figure 4.43 shows the areas on the beach that have been contaminated by oil and the locations of protective booms with spill forecasts during the month of May by NOAA. It should be noted that the reference for all dates without a year is 2010.

Researchers at Florida State University have also been systematically analyzing the radar images of this spill. Figure 4.44 shows the animated graphic with a detailed

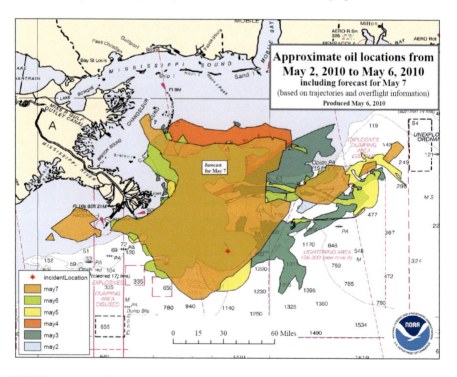

FIGURE 4.42 Spill locations on May 6 and 7 (NOAA/BBC).

Major Oil Spills

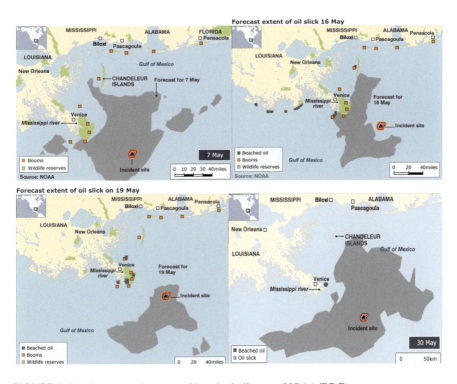

FIGURE 4.43 Protective booms and beached oil areas (NOAA/BBC).

FIGURE 4.44 Animation showing oil slicks moving eastward along the Alabama and Florida coasts. Image courtesy of Florida State University.

look at the northeastern portion of the oil slick as it moves eastward off the Alabama coast and the Florida Panhandle on May 31, June 1, and June 3.

Figure 4.45 shows the spread of oil as characterized by heavy, medium, and light oil. As the oil becomes lighter it can cover more area as shown in these forecasts by NOAA in the month of June, almost two months after the accident. A similar spill trajectory as was reported by NBC and simulated by NOAA is shown in Figure 4.45 as spill moving towards Florida beaches. Figure 4.46 shows an animation of oil trajectory for two months and 20 m below water as simulated by NCAR on June 3. Daily tracks of spill as reported by CNN for three months following the accident from April 26 to July 21 are shown in Figure 4.47.

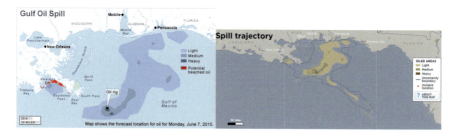

FIGURE 4.45 Areas covered by light, medium, and heavy oil on June 7 (source: NOAA).

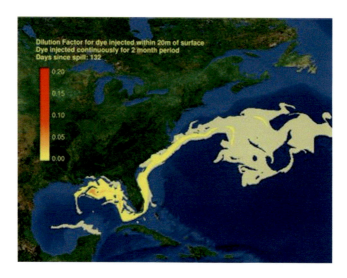

FIGURE 4.46 Animation of oil trajectory 132 days after the accident (NCAR, June 3).

Major Oil Spills

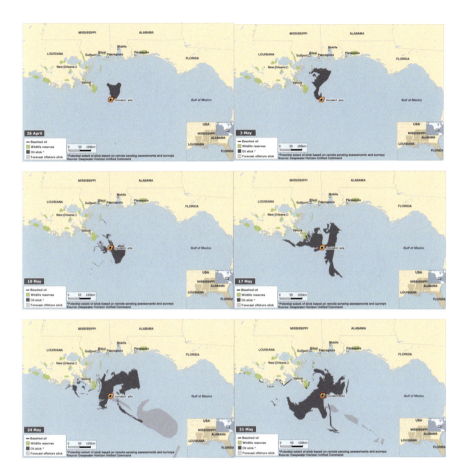

FIGURE 4.47 Simulated track of oil spill during April–July 2010 (source: Deepwater Horizon Unified Command/BBC).

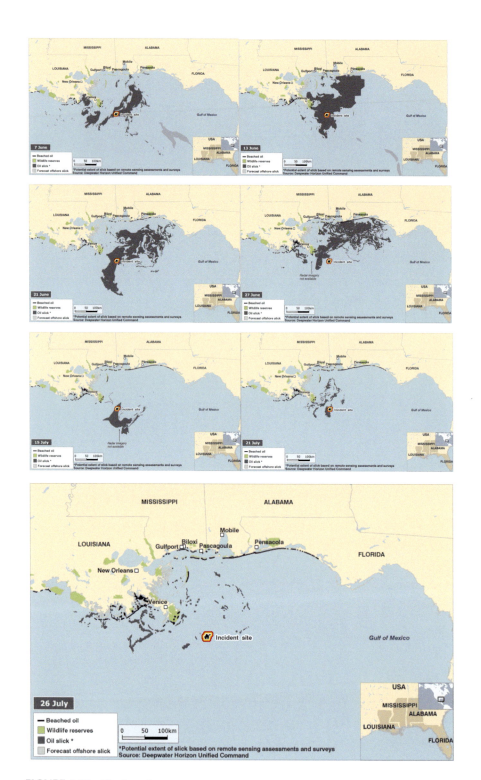

FIGURE 4.47 Continued.

Major Oil Spills

4.7 SUMMARY

In this chapter major oil spills occurring in recent history and from around the world were briefly reviewed. These major cases were caused by tanker collision/allision, war, or offshore production activities. For the case of the BP oil spill, as it happened most recently and was reported in the media extensively during 2010, it was covered in more detail in this and the following chapters. In Section 4.6, events as they occurred, the causes, lack of safety rules, and other factors that contributed to the worst environmental disaster in the US were discussed. Due to the significance of this event and availability of information, its economic and environmental impacts, response methods, and its modeling are covered in the following few chapters.

REFERENCES

Amato, E. 2003. *An environmental restoration programme 12 years after: The Haven Wreck, Les journees d'information du CEDRE.* October 6. https://wwz.cedre.fr/en/content/download/1730/16920/file/8-icram-12years-after-restoration-haven-1-en.pdf

AMSA. 2003. *Oil spill monitoring handbook.* Prepared by Wardrop Consulting and the Cawthron Institute for the Australian Maritime Safety, Canberra City, Australia.

Arab Times. 1997. English daily newspaper, Kuwait, July 15: 3. *Arab Times, 1998.* January 17, p. 7.

ASCE Task Force. 1996. State-of-the-art review of modeling transport and fate of oil spills. *Journal of Hydraulic Engineering*, 122(11), November, pp. 594–609.

BBC. 2010. Mexicans still haunted by 1979 Ixtoc spill, By Julian Miglierini. *BBC News*, June 14. https://www.bbc.com/news/10307105

BBC. 2018a. Sunken tanker Sanchi: Four oil slicks seen, says China. *BBC News*, January 6. https://www.bbc.com/news/world-asia-china-42728251

BBC. 2018b. *World news website.* January 9. https://www.bbc.com/news/world-asia-china-42615951

BBC World News. http://www.bbc.co.uk/news/world/ (accessed on September 17, 2010).

Bechtel. 2020. Taken from the following link on June 1, 2020. https://www.bechtel.com/projects/kuwait-reconstruction/

CBC. 2017. *Kuwait battles oil spill in Persian Gulf waters.* August 13. https://www.cbc.ca/news/world/kuwait-battles-oil-spill-in-persian-gulf-waters-1.4245702

Cedre. 2010. *Nowruz incident.* Updated May 18, 2010. http://wwz.cedre.fr/en/Resources/Spills/Spills/Nowruz

Cedre. 2014. *Exxon Valdez.* Updated March 25, 2014. Taken from the following link on May 25, 2020. https://wwz.cedre.fr/en/Resources/Spills/Spills/Exxon-Valdez

CNN. 2020. *Ten largest oil spills, CNN News website.* Updated February 26, 2020. https://www.cnn.com/2013/07/13/world/oil-spills-fast-facts/index.html

Deep Water. 2011. *The Gulf Oil disaster and the future of offshore drilling.* Report to the President, National Commission on the Deepwater Horizon BP Oil Spill, January.

EHSO. 2008. *Environment, health and safety.* http://www.ehso.com/oilspills.php, also see the EPA webpage on oil spills http://www.epa.gov/oilspill/

EIA. 2019. Today in Energy, US Energy Information Administration, June 20, 2019.

EMSA. 2013. *European Maritime safety agency.* As adopted by EMSA's Administrative Board at its 37th Meeting held in Lisbon, Portugal, on November 13–14 . p. 38.

Energy Global News. 1992. *The Fergana Valley massive oil spill.* March. http://www.energyglobalnews.com/march-1992-the-fergana-valley-massive-oil-spill/

Husseini, T. 2018. *The five biggest oil spills offshore: Lessons to learn.* Offshore Technology. August 10. https://www.offshore-technology.com/features/five-biggest-oil-spills-lessons/

ITOPF. 1989. *Exxon Valdez.* http://www.itopf.org/in-action/case-studies/case-study/exxon-valdez-alaska-united-stated-1989/

ITOPF. 1991. *HAVEN.* Italy. April 11. Taken from the following link on May 30, 2020. https://www.itopf.org/in-action/case-studies/case-study/haven-italy-1991/

ITOPF. 1995. *Braer, UK 1993.* January 5. https://www.itopf.org/in-action/case-studies/case-study/braer-uk-1993/

ITOPF. 2018. *Sanchi incident, ITOPF perspective.* http://www.pcs.gr.jp/doc/esymposium/2018/ITOPF_Mr_Alex_Hunt_ppt_E.pdf

ITOPF. 2019. *Oil tanker spill statistics 2018.* January. http://www.itopf.org/fileadmin/data/Documents/Company_Lit/Oil_Spill_Stats_2018.pdf; https://www.itopf.org/knowledge-resources/data-statistics/statistics/

ITOPF. 2020. International Tanker Owners Pollution Federation Limited, ITOPF, Ltd, London. Taken from the following link on May 5, 2020. www.itopf.org

Mansor, S.B., H. Assilzadeh and H.M. Ibrahim. 2006. Oil spill detection and monitoring from satellite image. *GIS Development 2006.* http://www.gisdevelopment.net/application/miscellaneous/misc027.htm

Montero, P., J. Blanco, J.M. Cabanas, J. Maneiro, Y. Pazos, A. Moroño, C.F. Balseiro, P. Carracedo, B. Gomez, E. Penabad, V. Pérez-Muñuzuri, F. Braunschweig, R. Fernandes, P.C. Leitao, R. Neves 2003. Oil Spill Monitoring and Forecasting on the Prestige-Nassau accident. In: Proc. Environment Canada's 26th Artic and Marine Oil spill (AMOP) Technical Seminar Otawa, Canada pp. 1013–1029.

Nguyen, P.Q.P. 2018. The oil spill incident in Vietnam. *European Journal of Engineering Research and Science, EJERS,* 3(7), July. https://pdfs.semanticscholar.org/960d/bfd32f5b311288544a19fe7c2e24b2b1bc35.pdf

NOAA – National Oceanic and Atmospheric Administration, US Department of Commerce. Washington, DC. http://response.restoration.noaa.gov/oil-and-chemical-spills (accessed on June 3, 2010).

Ordinioha, B. and S. Brisibe. 2012. The human health implications of crude oil spills in the Niger delta, Nigeria: An interpretation of published studies. *Nigerian Medical Journal: Journal of the Nigeria Medical Association,* December 31, 54(1), pp. 10–16.

PTJ. 2017. *Kuwait battles oil spill in the Persian Gulf by Klaus Ritt, Pipeline Technology Journal (PTJ).* August 21. https://www.pipeline-journal.net/news/kuwait-battles-oil-spill-persian-gulf

Reddy, C.M., J.S. Arey, J.S. Seewald, S.P. Sylva, K.L. Lemkau, R.K. Nelson, C.A. Carmichael, C.P. McIntyre, J. Fenwick, G.T. Ventura, B.A.S. Van Mooy and R. Camiili. 2011. *Composition and fate of gas and oil released to the water column during the Deepwater Horizon oil spill,* edited by J.M. Hayes. PNAS direct submission, www.pnas.org/cgi/doi/10.1073/pnas.1101242108; www.pnas.org/content/early/2011/07/15/1101242108.full.pdf, Also see supporting information by Reddy et al., 10.1073/pnas.1101242108; www.pnas.org/content/suppl/.../1101242108.../pnas.1101242108_SI.pdf

Riazi, M.R. 2012. Modeling of the rate of oil vaporization from GOM oil spill AIChE Annual Meeting, Pittsburg, PA October 28, 2012.

Riazi, M.R. 2010. *Accidental oil spills and control,* Chapter 5 in the book entitled "Environmentally Conscious Fossil Energy Production", edited by Myer Kutz and A Elkamel, John Wiley and Sons, New York.

Riazi, M.R. 2016. Modeling and predicting the rate of hydrocarbon vaporization from oil spills with continuous oil flow. *International Journal of Oil, Gas and Coal Technology (IJOGCT),* 11(1), pp. 93–105.

SCMP. 2018. East China Sea oil slick triples in size along whale migration route. *South China Morning Post*, January 22. https://www.scmp.com/news/china/policies-politics/article/2129985/east-china-sea-oil-slick-triples-size-along-whale

Sen, I. 2017. *Oil&Gas, Kuwaiti authority says oil spills contained, Oil&Gas Middle East.* August 23. https://www.oilandgasmiddleeast.com/article-17715-kuwaiti-authority-says-oil-spills-contained

Sykas, D. 2020. *Oil spills detection and identification with Synthetic Aperture Radar (SAR), Geo-University.* Taken from the following link on 10 May 2020, https://www.geo.university/pages/oil-spills-detection-and-identification-with-synthetic-aperture-radar-sar

Vogt, P. and D. Tarchi. 2004. *Monitoring of marine oil spills from SAR satellite data.* Ispra, Italy: European Commission. http://inforest.jrc.it/documents/2004/SPIE5569-3PVogt.pdf

Weather Channel Website. 2010. (www.weather.com) for Venice, Louisiana, May–September.

5 Environmental, Economic, and Political Impacts
Case of BP Oil Spill

ACRONYMS

AP   Associated Press
BP   British Petroleum
BOP   Blowout preventer
DWH   Deepwater Horizon
EPA   US Environmental Protection Agency
GOM   Gulf of Mexico
NCAR   US National Center for Atmospheric Research
NOAA   National Oceanic and Atmospheric Administration

5.1 INTRODUCTION

Every oil spill accident has its own and unique environmental and economic impacts on the society, industry, energy supply, and policy makers. The environmental and economic impacts also depend on the type of accident, location of the accident, amount of oil released, type and composition of oil, weather and sea conditions (temperature, air speed, surface water speed, water high, and waves, humidity, etc.), response methods, and the residents affected by the spill. For the case of the BP oil spill in which the leak was continuous for almost three months and the leak source was located 5,000 ft below the water surface but near residential and recreational areas, the situation is complicated in comparison with tanker collisions in the sea and not close to residential areas.

As the BP oil spill was the worst environmental disaster in US history releasing about 5 million barrels of oil into the Gulf of Mexico with significant consequences for the environment, economy, and offshore production plans, this chapter is devoted to a review of such impacts on the society, industry, and policy makers. In Chapter 4 (Section 4.6) we discussed how the BP oil spill occurred, the amount, expansion, and surface coverage, as well as properties of the reservoir fluid and characteristics of the Macondo well. Simulated and satellite images showed the areas covered by the oil and the shores affected. The US president visited the area four times during the time

135

of crisis and assigned a special commission to study the impacts of the spill. The crisis costs for the operating company were more than $60 billion, as the US government, US Congress, and US courts, along with the British government and public opinion, were all involved in shaping the events. This chapter describes the events as they developed during 2010. All dates cited without year refer to 2010.

5.2 ENVIRONMENTAL IMPACTS

5.2.1 Impacts on the Birds and Fish

In this part the effects of the oil spill on the life of sea birds, pelicans, fish, turtles, marine mammals, and gulls are shown mainly through the images that were published in 2010. In 2016 US government sources said more than 100,000 birds were killed as a result of the BP oil spill (US Fish and Wildlife Service, www.fws.gov). There were particular concerns about the lives of brown pelicans and piping glovers. Figure 5.1 shows oil-covered brown pelicans in the Gulf of Mexico near the Louisiana coast on June 3, 2010.

In just a six-week period prior to June 7, 292 birds were brought to the International Bird Rescue Research Center in Fort Jackson, Louisiana, of which 96 were brown pelicans. The birds were rescued and transported to the Fort Jackson Rehabilitation Center by well-trained and knowledgeable wildlife responders, veterinarians, and biologists (NCAR, June 7). The US Interior Secretary toured the center and watched bird cleaning by the volunteers as shown in Figures 5.2 and 5.3.

Figures 5.4 and 5.5 show how brown pelicans struggle against oil waves in water off the coast of Louisiana as covered by oil from Deepwater Horizon rig oil spill. Figure 5.6 shows a seabird dead as a result of the BP oil spill on Elmer's Island.

FIGURE 5.1 Oiled pelicans in the Gulf of Mexico, June 3, 2010 (source: coomns.wikimedia.org).

Environmental, Economic, and Political Impacts   137

FIGURE 5.2 A volunteer rescues a brown pelican at Fort Jackson Rehabilitation Center (NCAR, June 7) left.

FIGURE 5.3 Cleaning an oiled baby brown pelican chick at the BP 2010 Gulf oil spill response at Fort Jackson Bird Rehabilitation Center in Buras, Louisiana on June 22 (photo by International Bird Rescue Research Center (IBRRC)/Flickr) right.

Figure 5.7 shows some brown pelicans perched above oil-stained waters in Grand Isle, Louisiana, on June 19.

A pod of bottlenose dolphins swimming under the oily water in Chandeleur Sound, La., is shown in Figure 5.8 in May 2010 in the Gulf of Mexico. As the Gulf

138   Oil Spill Occurrence, Simulation, and Behavior

FIGURE 5.4 An oiled brown pelican struggles off the coast of Louisiana.

FIGURE 5.5 Birds struggling with oil from the Louisiana coast as a result of the BP oil spill (Flickr.com).

Environmental, Economic, and Political Impacts 139

FIGURE 5.6 A seabird dead as a result of BP oil spill.

FIGURE 5.7 Grand Isle, Louisiana – June 19, 2010: brown pelicans perch above oil-stained waters in the Gulf of Mexico.

oil disaster continued, BP was containing less than half of the daily leakage. Damage undersea was just as bad as on the oil-stained coastlines (CBS, June 13). On May 27 a new plume stretching 22 miles was found under the sea near Mobile, Alabama, as reported by scientists of Florida State University (CBS May 27).

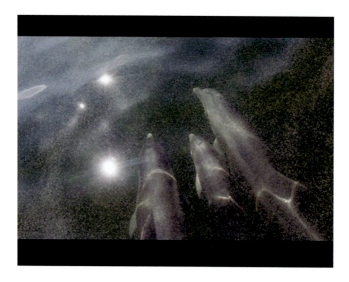

FIGURE 5.8 Dolphins swim under the oil water in Chandeleur Sound, La.

5.2.2 Impacts on the Shores and People

The Deepwater Horizon oil spill contaminated the shores of five states: Louisiana, Mississippi, Alabama, Florida, and Texas. By June 4 more than 200,000 km^2 of federal Gulf waters in these states (except Texas) were closed to fishing. The US shrimp and oyster supply is heavily concentrated in the Gulf area (Reuters, June 4). Figure 5.9 shows the simulated map and area covered by the spill on June 4. Figure 5.10 shows a dead fish on the Louisiana shore, a scene commonly seen in the spill-affected areas. Figures 5.10 and 5.11 show oil-covered crab on the beach of Grand Terrace Island, Louisiana.

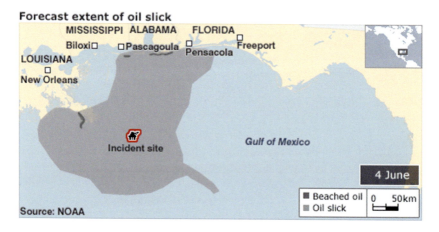

FIGURE 5.9 Simulated extent of the oil slick left (NOAA, BBC, June 4).

Environmental, Economic, and Political Impacts

FIGURE 5.10 This bottlenose dolphin was found dead in the Gulf of Mexico in July 2012. It may be one of more than 1,000 that appear to have died from poisoning related to oil spilled as a result of the 2010 Deepwater Horizon well blowout. (Louisiana Department of Wildlife and Fisheries)

FIGURE 5.11 An oil-covered crab stands on the beach of Grand Terre Island near Grand Isle, Louisiana, June 9, 2010 (source: UPI).

Figures 5.12 and 5.13 show movement of oil into the wetlands and marshes of Louisiana. The first heavy oil as it reaches the shores of Louisiana is shown in Figure 5.14.

On May 24, Louisiana Governor Jindal (Figure 5.14) expressed his frustration over BP and the government help as he was concerned about advancing oil spill into marshes and the shores. The oil had already hit 65 miles of shorelines (ABC, May 24). The governor said the oil moving into the wetlands was heavy oil (Figures 5.15). He was desperate to receive millions of feet in booms, more skimmers, more

FIGURE 5.12 An aerial view of Louisiana wetland inhabitants as oil moves into shorelines (photo by US Fish and Wildlife Service).

FIGURE 5.13 Crude oil in marsh grass, Barataria Bay, Louisiana (Shutterstock).

Environmental, Economic, and Political Impacts 143

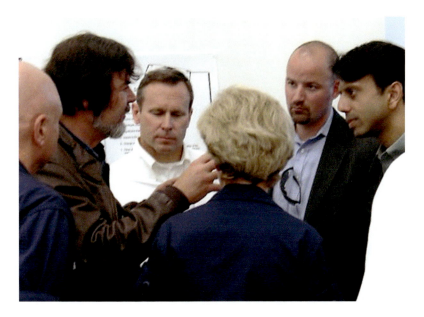

FIGURE 5.14 NOAA's Charlie Henry (second from left) discusses details of the response in the early days of the Deepwater Horizon oil spill with Louisiana Governor Bobby Jindal (right) and other key members of the response (source: NOAA).

FIGURE 5.15 The first heavy oil reaches the shores of Louisiana's marshlands (UPI/A.J. Sisco).

FIGURE 5.16 Seabird losses from the Deepwater Horizon oil spill estimated at hundreds of thousands (photo by Office of Governor Jindal/Louisiana/ScienceMag.org).

vacuums, more jack-up barges, and permission to build new barrier islands to prevent the oil from reaching marshes.

The Deepwater Horizon oil spill was the biggest environmental disaster in US history, and its impacts on marine life are usually underestimated. Birds are especially vulnerable to oil, which can coat their feathers and cause death by dehydration, starvation, or drowning (Figure 5.16). In Figure 5.17 a blue heron is shown standing next to a containment boom, and Figure 5.18 shows brown pelicans in mangroves in Breton Island in Louisiana. A more recent study showed that seabird mortalities from the BP oil spill in the Gulf of Mexico were estimated between 300,000

FIGURE 5.17 A blue heron next to a containment boom (photo by US National Park Service UPS/Flickr).

Environmental, Economic, and Political Impacts 145

FIGURE 5.18 Brown pelicans in mangroves on Breton Island, La. (USFWS photo).

to 2 million with a likely number of around 700,000 deaths, which is equivalent to more than 30% of the population in the GOM area. During the *Exxon Valdez* oil spill, the seabird deaths were estimated between 160,000 and 320,000 (Jennifer Balmer, October 31, 2014, www.ScienceMag.org). Figure 5.19 shows a long dolphin washed ashore dead in May 2011. Dolphin deaths reported in the Gulf of Mexico increased dramatically in the decade following the BP oil spill (VOAnews, Matt Haines, August 8, 2020).

On June 5 the oil from the Deepwater Horizon rig leak started washing ashore on the Alabama and Florida coast beaches. Figure 5.20 shows heavy amounts of

FIGURE 5.19 An 8-ft-long dolphin dead in May 2011 (credit: Healthy Gulf).

FIGURE 5.20 Deepwater Horizon-related oil on Orange Beach, Alabama, June 24, 2010 (source: www.researchgate.net).

oil making their way on the tide to shore in Orange Beach, Alabama, where a large sheen of oil coated much of the coast (CBS June 12). Oil spills can damage the air as badly as on the land and water surface, especially for light oils which contain significant amounts of light compounds that easily vaporize. On May 28 and on day 37 of the Gulf spill, ABC News reported that nine fishermen in the Gulf suffered from severe nausea, headaches, and trouble breathing, requiring the entire fleet of 125 boats in the area to evacuate and return to the shore for medical evaluation. Four fishermen were taken to local hospitals. These symptoms are consistent with oil or volatile organic compound gas exposure (ABC, May 31). One of the key compounds in oil is benzene which is very toxic and hazardous to the environment both in the air and in the water. Excess levels of benzene in air or water is toxic to humans. According to the United States Occupational Safety and Health Administration (OSHA), benzene is a cause of cancer in humans, and toxic levels have been linked to an increased risk of leukemia and heart and nervous system problems. OSHA has set the benzene permissible exposure limit at 1 ppm and a short-term exposure limit at 5 ppm for 15 minutes. The Centers for Disease Control and Prevention (CDC) reports that breathing in benzene at high levels can lead to drowsiness, irregular heartbeats, and headaches, and with continued exposure or very high levels could result in unconsciousness or even death. Equipment used to measure levels of benzene and aromatics in the air is simple and handy as shown in Figure 5.21.

Environmental, Economic, and Political Impacts 147

FIGURE 5.21 An equipment to measure benzene and total aromatics in the air (source: https://www.ionscience.com/).

Light components are also more soluble in water. In addition, dispersants used to remove oil from the water surface are also toxic. Divers going underwater to monitor the spill under the water needed special protective equipment and waterproof dive suits to protect themselves from contamination by such compounds. Some benzene reaching the water surface vaporizes into the air but as it is heavier than air it can accumulate close to the water surface, further increasing exposure risk to workers at sea during cleanup operations (ABC, May 31). Health officials reported respiratory and skin irritation problems in Louisiana and Alabama. Louisiana Department of Health and Hospitals reported 71 cases of oil spill-related illness, of which 21 were from the general public (CNN, June 9). States were tracking the health consequences of the BP oil disaster in the Gulf of Mexico, including respiratory and skin irritation problems in Louisiana and Alabama, health officials said. Symptoms reported by workers included throat irritation, cough, chest pain, headaches, and shortness of breath. Figure 5.22 shows how the workers may come into contact with oil from the oil spill during cleanup operations.

FIGURE 5.22 Health, safety, and environment (HSE) workers contracted by BP load oily waste onto a trailer on Elmer's Island, just west of Grand Isle, La., May 21, 2010 (USCG photo by Patrick Kelley).

FIGURE 5.23 A C-130 Hercules from the US Air Force drops an oil-dispersing chemical into the Gulf of Mexico on May 5, 2010 as part of the Deepwater Horizon Response effort (photo by US Air Force/Sgt. Adrain Cadiz).

Environmental, Economic, and Political Impacts 149

FIGURE 5.24 A US Coast Guard vessel in waters filled with the toxic chemical Corexit (source: https://commons.bluemountaincenter.org/).

Another source of toxic materials hazardous to the marine life and human body after an oil spill occurs is dispersants which are chemicals that are sprayed on the surface of an oil slick to break down the oil into smaller droplets that can be easily mixed with the water (Figure 5.23). Chemical dispersants were sprayed in unprecedented quantities in the Gulf as part of the response to the BP oil spill. Figure 5.23 shows an aircraft from the US Air Force spraying a chemical dispersant on the Gulf of Mexico on May 5. One of the dispersants used at the BP oil spill, Corexit 9527A, contains 2-Butoxyethanol which may cause injury to red blood cells (hemolysis), kidney, or the liver according to the manufacturer's safety data sheet (www.biologicaldiversity.org). Figure 5.24 shows a US Coast Guard vessel in the waters of the Gulf of Mexico with toxic chemical Corexit.

5.3 ECONOMIC IMPACTS

The BP oil spill had great impacts on the economy of five states as well as enormous cost for the company. BP was the operator and principal developer of the Macondo Prospect with a 65% share, while 25% was owned by Anadarko Petroleum and 10% by MOEX Offshore 2007, a unit of Mitsui (CNN, December 16). The Obama administration sent a $69 billion bill to BP for the government's efforts to help deal with the spill (CNN, June 4).

Five years after millions of gallons of oil flooded into the Gulf of Mexico, BP agreed to pay the federal government and five Gulf states a record $18.7 billion for damage caused by the spill. It was referred to as the largest environmental settlement in history. BP agreed to pay $7.1 billion to Louisiana, Alabama, Florida,

Mississippi, Texas, and the federal government over the course of 15 years for the environmental damages. It also agreed to pay a $5.5 billion civil penalty to the US government under the Clean Water Act, $4.9 billion to five Gulf states to settle damages, and $1 billion to several hundred local government entities (The Atlantic, July 2015). BP wrote more than 75,000 checks for the claims it received, and its staff dedicated 74,000 hours to the claims (CNN, July 21). At the time of the accident BP had more than 10,000 employees just in the UK. However, after six weeks from the accident, one-third of BP value was wiped out (CNN, June 4). Within three months from April to July 2010 the value of BP shares reduced by half as shown in Figure 5.25.

In addition to the government fines, BP paid all types of people including fishermen in Louisiana, motels in Mississippi, and schools in Florida as well as compensation for medical costs, property damage, economic losses, and its own cleanup. Although the Deepwater Horizon rig was owned by Transocean, it was leased by BP and BPwas responsible for cleanup of the rig (Washington Post, July 14, 2016). On December 16, 2010, BP sued Transocean, the owner of the Deepwater Horizon rig, for $40 billion in damages (CNN, Dec. 16).

Many fishermen were put out of work by the oil spill and took jobs in the cleanup efforts. Fishermen in North Carolina, South Carolina, Georgia, and Texas, whose waters had not been affected by oil, said prices for their shrimp had gone up as processing plants that normally bought Gulf seafood turned to other docks for their supply. Because of the oil spill about a third of federal waters in the Gulf were closed to fishing boats for fear of contaminated seafood. In one

FIGURE 5.25 BP shares lost more half of their value in just three months from April to July 2010 (source: commons.wikimedia.org).

Environmental, Economic, and Political Impacts

shrimp town in the area prices went up by 30%. The federal government declared fishery disasters for Louisiana, Mississippi, Alabama, and Florida, which brought emergency payments for commercial fishermen (ABC, June 12). By June, the commercial and recreational fishing closure reached 60,683 square miles, which is about 25% of the Gulf of Mexico exclusive economic zone, according to the National Oceanic and Atmospheric Administration. Images from the massive BP oil spill prompted tourists to go to other destinations (ABC, June 7). In total, BP's big bill for the world's largest oil spill reached $61.6 billion (Washington Post, July 14, 2016).

5.4 HOW THE OIL SPILL WAS STOPPED

Attempts to stop the flow of oil into the sea began right after the accident happened on April 20, first by trying to extinguish the fire on board and after the rig sank on April 22. The surroundings of the Deepwater Horizon accident are shown in Figure 5.26 with 24 skimming vessels, 20 support vessels, and 3 drilling rigs in the Gulf of Mexico. BP officials tried to use various techniques to contain the flow of oil. After failing to close the BOP valves, on May 6 a large containment vessel was used to contain the oil (Figure 5.27).

The BP plan was to use a 40 ft tall iron box weighing 98 tons to collect oil from 5000 ft down on the seabed as demonstrated in Figure 5.28. BP hoped to be able to collect 85% of the leaking oil and pipe it to the surface (BBC, May 6). For this purpose, a ship reached the site to collect oil. This method, however, had been used

FIGURE 5.26 Surrounding the site of the Deepwater Horizon incident (courtesy: US Coast Guard/*The Chemical Engineer Journal*).

FIGURE 5.27 A huge containment vessel was used to contain the DWH oil on May 6 (photo by US Coast Gulf).

FIGURE 5.28 Collecting oil from the leak with use of a funnel (image: USCG/BP).

to contain oil on the surface but was never tried at 5000 ft below the water surface. This method did not work because of the formation of gas hydrate due to the presence of hydrocarbon gases and cold water in the vessel. The third failed attempt was the insertion of a long narrow tube into the broken pipe using underwater robots. It

Environmental, Economic, and Political Impacts   153

FIGURE 5.29 Gas from damaged DWH wellhead is burned by *Discoverer Enterprise* on May 16 (photo by USCG/Patrick Kelley).

FIGURE 5.30 On May 26 BP tried its top kill procedure to stop the flow of oil by placing heavy mud into the oil well. This photo is a frame grab of the live video stream of operations (UPI/BP).

was thought that the 6-in tube and stopper could capture more than three-quarters of the leak (BBC May 17). Gas from the damaged wellhead was burned by the drillship *Discoverer Enterprise* in a process known as flaring (Figure 5.29). Oil and gas from the wellhead were brought to the surface via a tube that was placed inside the damaged pipe. On May 26, BP executed another method known as the "top kill" process, which places heavy kill mud into the oil well in order to reduce pressure and the flow of oil from the well as shown in Figure 5.30. The procedure was not successful because the pressure from the oil and gas was very high and the junk mud

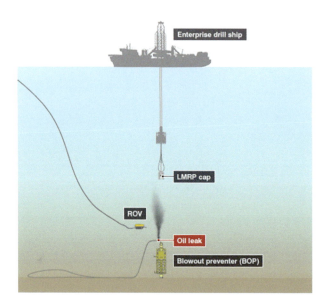

FIGURE 5.31 In June BP attempted a new method to put a cap on the BOP (graphic of BP/ reported by BBC, June 4).

was not able to stop the flow of oil. In the next move, BP attempted to siphon oil from the top of the blowout preventer. The damaged riser, which is the pipe taking oil from the well, was cut using a remotely operated shear. The plan was to lower a cap onto the upper section of the blowout preventer known as the lower marine riser package (LMRP). BP announced that oil and gas was being received on board the *Discoverer Enterprise* following successful placement of a cap on top of the failed BOP as shown in Figure 5.31 (BP/BBC, June 4).

Finally, on July 15 at 2:25 pm Houston time, BP announced that it had successfully plugged the oil leak using a larger and tighter cap than previously failed caps. For the first time, 86 days after the DWH rig explosion, the flow of oil was completely stopped and the company was hoping to collect as much as 28,000 barrels per day. However, to seal the well permanently, the well had to be sealed off by cement as shown in Figure 5.32a. In the method known as static kill, first the company injected 2300 barrels of heavy drilling mud that pushed oil back into the reservoir. The rate of injection was 5 barrels per minute which increased to 10 and then 15 barrels a minute near the end of the operation which took 8 hours (CNN, August 4). As part of this process two relief wells were drilled (Figure 5.32b). The process took almost four months, and finally BP was successful in permanently sealing through closing the top of the well by cement. A US official who was supervising the operations announced on Saturday, September 18, 2010, that the Macondo well had been permanently sealed off and it was called a dead well.

Environmental, Economic, and Political Impacts

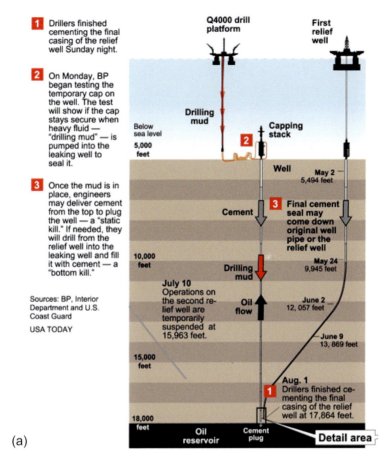

FIGURE 5.32 (a) Permanent sealing of the well with cement (August 2, USA Today/BP/DOI/USCG).

5.5 BRIEF HISTORY OF BP

BP first was born as the Anglo-Persian Oil Company in 1909, one year after oil was found in southern Iran. It built the Abadan refinery which was the largest in the world during the first half of the 20th century with a capacity of more than half a million barrels per day. The British government became the company's major stakeholder on the eve of World War I. During World War II, the Abadan refinery continued to feed the Allied war machine. By the end of World WarII, says BP's own website, "war without oil would be unimaginable."

In 1954, after nationalization of the Iranian oil industry, the company branded itself as British Petroleum Company Limited or **BP**. After the Iranian revolution, the historic Abadan refinery was destroyed by Iraqi forces in 1980. After merging with Amoco in 1998, the corporation took the name BP-Amoco before assuming

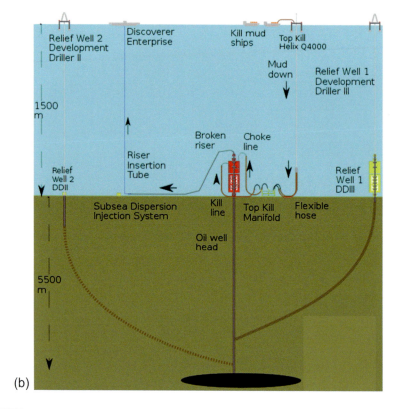

FIGURE 5.32 (b) Sealing of the Deepwater Horizon wellhead (source: Wikimedia).

the name BP PLC in 2000. The company's headquarters are in London. In 2001 the company branded itself again as "Beyond Petroleum" with a new and greener sign as shown in Figure 5.33. Its logo, a flowery pastel helix, beams earthy friendliness while the company's current tagline – "Beyond Petroleum" – expresses its desire to diversify into sustainable and greener forms of energy (Time, June 14, 2010, p. 18).

In 2006 BP experienced a massive oil spill in Prudhoe Bay, Alaska. It was alleged that cost-cutting measures by BP executives led to poor maintenance which caused

FIGURE 5.33 The left is the old sign and the right is the current and new sign of BP (source: Wikipedia).

Environmental, Economic, and Political Impacts 157

FIGURE 5.34 President Barack Obama and Vice President Joe Biden meet with BP executives in the Roosevelt Room of the White House, June 16, 2010, to discuss the BP oil spill in the Gulf of Mexico. Pictured, from left, are BP CEO Tony Hayward, BP Chairman Carl-Henric Svanberg, BP General Counsel Rupert Bondy, BP Managing Director Robert Dudley, Senior Advisor Valerie Jarrett, Labor Secretary Hilda Solis, Attorney General Eric Holder, Biden, Obama, and Homeland Security Secretary (official White House photo by Pete Souza).

the accident. The 2010 GOM oil spill was the worst accident in the company's history. The disaster cost the company more than $60 billion, and from April to July 2010, within three months, its share price dropped by more than 50%. After the accident, on July 28, the company announced that Tony Hayward would leave his post as CEO by October 1 and a new CEO would take over. Figure 5.34 shows BP officials meeting the US president and other officials at the White House to discuss the BP oil spill on June 16.

5.6 POLITICAL IMPACTS

The 2010 BP oil spill in the Gulf of Mexico had great impacts within the US and far beyond its borders. This environmental disaster was one of the early challenges in the Obama presidency and the president was heavily involved in the planning of the response strategy to the disaster. Obama made four trips to the region affected by the oil spill. Figure 5.35 shows him talking with a local official while visiting a beach in Louisiana. Vice President Joe Biden also traveled to Theodore, Alabama, to assess efforts to counter the BP oil spill. During the visit, the vice president toured the Theodore Staging Facility and met with fishermen and small business owners from the area. He was joined by National Incident Commander Admiral Thad Allen and NOAA Administrator Dr. Jane Lubchenco as shown in Figure 5.36 (obamawhitehouse.archives.gov).

FIGURE 5.35 President Obama visits the region with Lafourche Parish President Charlotte Randolf (left) and inspects a tar ball as they examine the impact of the BP oil spill on Fourchon Beach in Port Fourchon, La., May 28, 2010 (official White House photo by Chuck Kennedy via White House/Flickr).

FIGURE 5.36 Vice President Joe Biden is led on a tour of the boom repair area by Walt Dorn of Patriot Environmental Services, at the Theodore Staging Area in Alabama, July 22, 2010 (obamawhitehouse.archives.gov).

Environmental, Economic, and Political Impacts

CBS conducted a poll about one month after the accident. About 70% disapproved of the way BP was handling the crisis just five weeks after the accident while the approval rate was at 18% (CBS, May 25). The Obama administration received 45% disapproval and 35% approval for their handling of the BP oil spill by May 25. Several polls were conducted later by ABC News/Washington Post on June 7 which showed that the people viewed the BP oil spill as worse than Hurricane Katrina in August 2005. Most people also favored criminal charges against BP and others involved in the Gulf of Mexico oil spill as they considered it a major environmental disaster. According to the ABC/Washington Post poll conducted on June 7, 81% had a negative view of BP's response while 69% had a negative view of the federal government's response to the disaster. According to the same poll on May 9 only 55% considered the BP oil spill as a major disaster while this number reached 73% a month later on June 6 (ABC, June 7). BP officials were under pressure from federal authorities to contain the massive oil spill, especially as the new estimates of the amount of oil were on the rise. In addition, environmentalists were saying demonstrations against BP's response to the disaster were unfolding in more than 50 cities across five continents, from Pensacola, Florida, to Christchurch, New Zealand (CNN, June 12). On June 13, the UK prime minister and the US president spent 30 min on the phone discussing the oil spill and how to bring it under control as well as its impact on the relation between the two countries (BBC, June 13).

In another trip, Obama visited the Theodore Staging Facility in Theodore, Alabama, where containment booms were prepared on Monday, June 14, 2010. Obama said the oil spill accident was on the same scale as 9/11 from a national tragedy point of view and its importance (BBC Radio, June 14). One day after his return from Alabama, he made his first Oval Office address to the nation (Figure

FIGURE 5.37 President Obama addressing the nation on the BP oil spill from the Oval Office on June 15 (source: obamawhitehouse.archives.gov/June 15, 2010).

5.37), detailing the oil spill response and calling for a new clean energy policy to end US dependence on fossil fuels (ABC, June 15). In addition to President Obama, Vice President Biden also visited Louisiana and the oil-fouled Florida Panhandle (CNN, June 26).

On many occasions the executive chairmen of BP and other related companies appeared in the White House and the US Congress in Washington, DC during May and June 2010. The chairmen of BP America, Transocean, and Halliburton were sworn in before a Senate Energy hearing on the BP oil spill accident in Washington, DC on May 11, 2010. US lawmakers had two days of hearings in Washington on the deadly drilling rig explosion and the oil spill that threatened an economic and ecological catastrophe on US Gulf shores (NBC News May 11). BP CEO Tony Hayward appeared before the US Congress on June 17 where he was grilled for more than seven hours (NBC News). Protesters against the BP executive greeted Hayward in the US Congress (Figures 5.38 and 5.39). Demonstrators against BP in Washington, DC on June 4 are shown in Figure 5.40. On September 15, 2010, Hayward appeared before the Energy and Climate Committee in the UK House of Commons in London. He gave some evidence to the committee studying the fallout of the Gulf of Mexico oil spill and the future of deep-water drilling (Parliament, UK and BP). On October 1, 2010, American Robert Dudley became BP's first non-British chief executive.

On May 21 President Obama formed a seven-member bipartisan commission to study and provide recommendations on the BP Gulf of Mexico oil spill and offshore drilling activity in the gulf area. The committee prepared its final report on January 11 titled: *Deep Water, The Gulf Oil Disaster and the Future of Offshore*

FIGURE 5.38 BP CEO under fire at Congress before the committee on Energy and Commerce on June 17 (Flickr.com).

Environmental, Economic, and Political Impacts 161

FIGURE 5.39 An activist protesting against BP while its CEO appears before the US Congress in June 2010 (Greenpeace/Flickr).

FIGURE 5.40 Demonstrators protest the BP oil spill, demand environmental justice, June 4, 2010, in Washington, DC (image: Shutterstock ID: 54543442).

Drilling, Report to the President, National Commission on the Deepwater Horizon BP Oil Spill, January 2011. Although official work dealing with the disaster ended, the impacts of the accident on US and perhaps world energy policy and related regulations remained for many years after the tragedy.

5.7 SUMMARY

The BP or Deepwater Horizon oil spill introduced in the previous chapter was the biggest oil spill with more information available in public domain than any other accident in history with significant consequences as it was considered worse than Hurricane Katrina and on the same scale as the 9/11 tragedy. In this chapter the impacts of the oil spill on the environment, policy, and decision makers were reviewed. It affected the five states of Louisiana, Alabama, Tennessee, Florida, and to a lesser extent Texas. The president and vice president toured the area five times, and a White House commission was formed to make recommendations about future policy regarding offshore drilling activities. Oil spill impacts on the economy of the region as well as on the operating company BP were also reviewed. Several attempts were made by BP to stop the flow of oil, until mid-July after 86 days the flow was stopped and in September the well was permanently sealed off by top kill method. Overall the disaster cost BP over $60 billion including fines and cleanup operations, causing its share price to drop by more than 50% within three months of the accident. A brief history of BP was also presented from the date of its registration in London in 1909.

6 Oil Spill Response and Cleanup Methods
Case of BP Oil Spill

ACRONYMS

AP	Associated Press
BP	British Petroleum
DWH	Deepwater Horizon
EPA	US Environmental Protection Agency
EMSA	European Maritime Safety Agency
GOM	Gulf of Mexico
ITOPF	International Tanker Owners Pollution Federation
NCAR	US National Center for Atmospheric Research
NOAA	National Oceanic and Atmospheric Administration
USGS	US Geological Survey

6.1 INTRODUCTION

In Chapter 5 the impacts of oil spills on the environment and related economic damages were discussed for the case of the BP oil spill. As soon as an oil spill occurs, attempts begin to minimize its impact and eventually to clean the area completely from the oil. However, for the case of Deepwater Horizon spill, the oil was flowing 5000 ft below the sea surface at a rate of 57,000 b/d, and in addition to cleanup at the surface, oil was traveling from the sea bottom to the sea surface, and there was continuous oil pollution in the water below the surface as well. Stopping the flow of oil took nearly three months as discussed in the previous chapter, so as oil was removed or evaporated from the surface, at the same time fresh oil was leaking into the water. Before choosing any response technique it is important to understand the natural processes by which oil could be partially vaporized, dispersed, dissolved, or sedimented without any action. These processes are discussed at the beginning, and their modeling and predictions are discussed in Chapters 7 and 8 following this chapter.

Man-made cleaning techniques to remove oil from the sea include the use of booms, dispersants, controlled burning, skimming, sorbent materials, low-pressure flush, vacuum cleaning, the mechanical removal of oil such as the use of shovels, bioremediation, solidification, etc. Manual or mechanical cleanup methods consist of placing workers on the coast, armed with shovels, rakes, and gloves to collect oil that has run ashore. Mechanical cleanup requires heavy machinery and is used in areas that are plagued by heavy oiled beaches or areas which are covered in debris. Sand berms are used to prevent oil reaching onshore.

Before deciding how to clean an oil spill, it is important to simulate the spill to determine how much oil would go through the natural processes versus time. In-situ burning itself could contribute to air pollution, and the use of dispersants could threaten marine life. In some areas natural cleaning could outweigh the benefits of cleaning by use of equipment. For example, wave actions, sunlight, and natural water dispersion contribute to break down oil leaked into the sea. Although for the people it would be difficult to understand why the crew wait and do nothing, sometimes natural recovery is the best option (NBC, June 9). However, the rate at which these processes take place is a very important factor. In the case of light oils and in hot areas, oil can be quickly vaporized, so the knowledge of the rate and amount of vaporization is vital before deciding on a cleaning method. BP opened a website (deepwaterhorizonresponse.com) and a phone line inviting the public to propose suggestions on the cleanup methods. For example one idea was suggested by a company called Clean Kool to use carbon dioxide to freeze the oil, making it easy to clean up (CBS May 6). In this chapter all these methods are reviewed especially for the case of the BP oil spill cleaning which was one of the most expensive cleanup operations in US history with a cost of $25 million per day (CNN, June 24).

At the first stage for the case of the BP oil spill the response crews battled the blazing remnants of the Deepwater Horizon rig on Wednesday April 21, one day after the accident as shown in Figure 6.1 (USCG).

On July 29 CNN reported the following data on the response activity:

- An estimated 3–5.2 million barrels discharged into the Gulf
- 827,046 barrels collected or flared
- 265,450 barrels flared in controlled surface burns

FIGURE 6.1 Fire boat response crews battle the blazing remnants of the offshore oil rig Deepwater Horizon, April 21, 2010 (US Coast Guard photo).

Oil Spill Response and Cleanup Methods

- 825,525 barrels of oily liquid skimmed
- 1,072,514 gallons of dispersants used

Deployment

- 24,800 people
- 4,300 vessels
- 800 skimmers
- 114 aircraft
- 3.41 million feet of boom deployed

By the end of July, BP had spent more than $4 billion on the response alone (CNN July 29).

By June 12, part of the BP oil spill reached the beaches of Florida (Figure 6.2). In this chapter, similar to previous chapters, all dates without a year refer to the year 2010.

6.2 NATURAL PROCESSES

Natural processes as shown in Figure 6.3 are the most important cleaning method. These processes are spontaneous without any human intervention required, as discussed in this section.

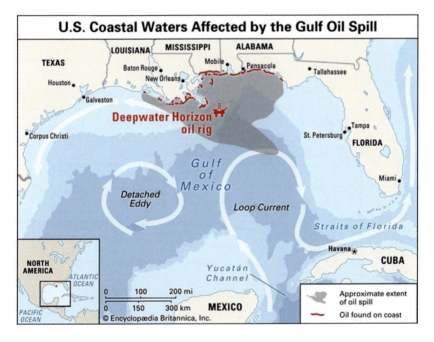

FIGURE 6.2 Deepwater Horizon oil spill of 2010: path of the oil three months after the accident (Britannica April 13, 2020).

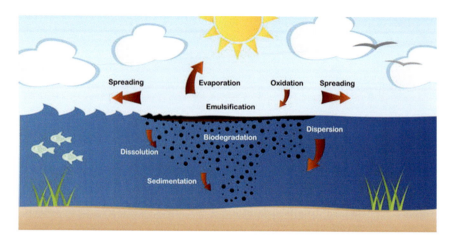

FIGURE 6.3 Major natural oil transport and weathering process (ITOFP, 2018).

6.2.1 Evaporation

Light compounds with lower boiling points can vaporize at lower temperatures, and as the temperature of oil increases heavier compounds may also vaporize. The main source of heat to vaporize the oil is from the sun. The amount of oil which can be vaporized depends on the temperature, composition, and properties of the oil, the surface area of the slick, and the wind speed, and varies from 20 to 60% of crude oil. As will be shown in Chapter 7, the rate of vaporization can be calculated once these data are known. For very volatile oil or petroleum products the oil maybe vaporized completely and quickly. The rates of evaporation of different types of oils are shown in Figure 6.4. The remaining oil after evaporation becomes heavier and thicker and less likely to dissolve naturally. Approximately the rate and the amount of natural disappearance of oil from different processes are compared and demonstrated in Figure 6.5. The width of each band indicates the importance of the process.

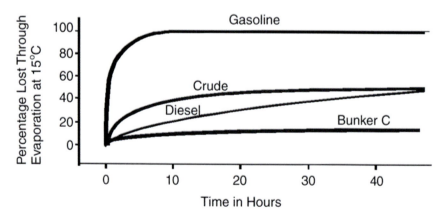

FIGURE 6.4 Evaporation rates of different types of oil and petroleum products at 15°C (NRC, 2003).

Oil Spill Response and Cleanup Methods

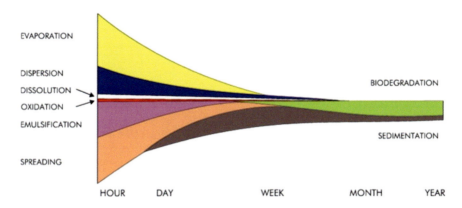

FIGURE 6.5 Rates and relative amounts for oil disappearance with time by different natural processes (Barrios, 2016/SINTEF).

6.2.2 Spreading

The oil particles when reaching the surface of water begin to spread, and the process is called spreading and it is important in the early hours as shown in Figure 6.5. After a few hours, under the influence of currents and waves, the rate of spreading reduces. The most important oil properties that determine the rate of spreading are its surface tension and viscosity derived by temperature as discussed in Chapter 2. Waves and turbulence contribute to the break-up of the slick into smaller particles as shown in Figure 6.6. Oils with lower surface tension have also lower interfacial tension between oil and water and can spread faster than heavier oils. In addition, parameters such as the amount of oil, water current, and temperature also have strong effects on the rate of spreading. At higher temperatures and warmer weather, the surface tension and density of oil reduce and it spreads faster. The spreading oil may reflect light in different directions, creating gray, metallic, or rainbow sheens (Figure 6.7).

6.2.3 Dispersion

Dispersion happens when weathered oil breaks down into smaller particles and these particles sink below the water surface and mix with the upper water column. Surface conditions may cause the oil slick to break into smaller droplets and mix with water. Some larger particles rise back to the surface and form a thin film known as sheen. Oil sheen is a very thin layer of oil usually with a thickness of 0.003 mm or less as shown in Figure 6.8.

6.2.4 Dissolution

Dissolution is a physical process by which, when oil is in contact with water, it may dissolve in the water. Generally the solubility of oil in water depends on the water temperature and composition of the oil and the conditions of the sea. Smaller and

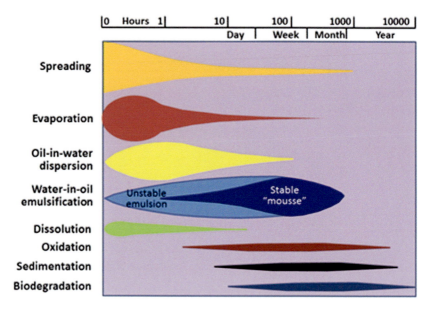

FIGURE 6.6 Spreading and evaporation are important in the early hours of a typical oil spill (courtesy of SINTEF).

FIGURE 6.7 Waves and turbulence at the sea surface can cause the break-up of the slick into small droplets (ITOPF).

lighter components have higher solubility than larger and heavier molecules. In addition, with an increase in temperature, the solubility of oil in water also increases as will be discussed in Chapter 7. Although the amount of dissolution of oil in water is small in comparison with the rate of evaporation, from a toxicological point of view it is important to know how much oil is dissolved in water, especially the aromatic portion of oil.

Oil Spill Response and Cleanup Methods

FIGURE 6.8 An oil sheen from BP oil spill on May 5, 2010 (source: DOI photo).

6.2.5 Emulsification

Emulsification is the mixing of oil in water, and it is caused by waves in the sea or ocean. This process can increase the volume of oil to four times its initial volume. Emulsification is resistant to other weathering processes, and for this reason when emulsification occurs more oil will stay on the water surface. Water-in-oil emulsions are commonly known as "chocolate mousse" because of their reddish-brown color and foamy texture (NBC, June 9). When the oil mixes with water, it changes from black to reddish-brown to orange. The oil typically sorted into long but relatively narrow strands of thicker oil, with broad areas of sheen as shown in Figure 6.9 (NOAA).

6.2.6 Oxidation

The reaction of oil with oxygen is called oxidation, and this process may happen when oil is floating on the water surface. Oxidation either breaks up the oil or creates tars such as tar balls which are the size of a coin and last a very long time after the spill. The process is shown in Figure 6.10.

6.2.7 Sedimentation

When oil breaks up into particles of different size those particles that are heavier than water may fall to the bottom of sea and smaller particles may be washed up on shorelines. In the sedimentation process, oil particles may attach to the sediment and sink to the ocean's floor. Chemical dispersants promote break-up and sedimentation

FIGURE 6.9 Emulsified oil on the sea surface as seen from an airplane on May 18, 2010 (source: NOAA).

process as well. Factors that affect the sedimentation of an oil particle are demonstrated in Figure 6.11.

6.2.8 Biodegradation

In an aquatic environment, microorganisms such as bacteria, fungi, and algae feed on oil and break it into smaller particles. The process is very slow; however, the rate of biodegradation depends on the amount of microorganisms, oxygen, temperature, and the size of oil particles. Biodegradation may affect smaller particles of oil which may stay in water for a long period of time. Warm water facilitates more

FIGURE 6.10 The oxidation process of oil spill and formation of tar balls (NOAA).

Oil Spill Response and Cleanup Methods 171

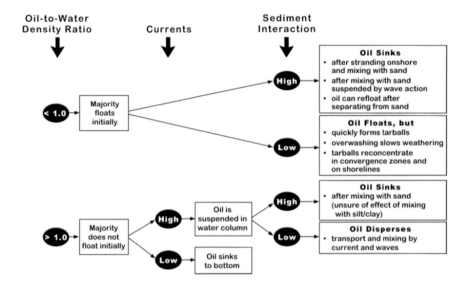

FIGURE 6.11 Factors determining whether spilled oil will float or sink (source: NIH/NRC, 2003).

biodegradation. Figure 6.12 shows how Curtis Franklin, an environmentalist, measures water salinity, temperature, and PH in the Gulf on May 12, 2010. As shown in Figures 6.5 and 6.6, the amount of oil biodegradation, oxidation, and sedimentation is small in comparison with evaporation, spreading, and dispersion.

6.3 PROTECTIVE BOOMS

Booms are floating physical barriers designed to perform one or more of the following functions:

FIGURE 6.12 Water testing in the Gulf in response to the BP oil spill on May 12 (photo by EPA).

- Oil containment and concentration: to prevent oil spreading over water surface and to increase its thickness to facilitate recovery.
- Deflection: diverting the oil into a suitable collection point on the shoreline for subsequent removal by vacuum trucks, pumps, etc.
- Protection: diverting the oil from economically important or biologically sensitive sites such as harbor entrances, power stations, etc.

Booms cannot remove oil spills as in-situ burning can, but rather they prevent oil from reaching certain areas and concentrate it in some locations for either burning or skimming. The most important characteristics of a boom are its oil containment and deflection capabilities. They are made of a buoyant material and longitudinal tension member (chain or wire) to withstand forces from winds and waves (ITOPF). There are several types of booms, but the most commonly used are hard and sorbent booms. A hard boom is like a plastic piece that has a cylindrical float. A sorbent boom looks like a long sausage made out of a material that absorbs oil. Sorbent booms don't have the "skirt" that hard booms have, so they can't contain oil for very long (NOAA). Waves, currents, tides, and rapid spreading of the oil could limit the operation of booms. Effective boom design and a well-planned and coordinated response can reduce these problems, although in some circumstances the use of any boom might be inappropriate (ITOPF). An NOAA simulation of the BP oil spill and location of booms two weeks after the accident is shown in Figure 6.13. Figure 6.14 shows how an inflammable boom is deployed in a U configuration between two vessels to contain heavy crude oil. Recovery of the oil will bring the cleanup operation to a successful conclusion (ITOFP). The US Navy was also actively involved

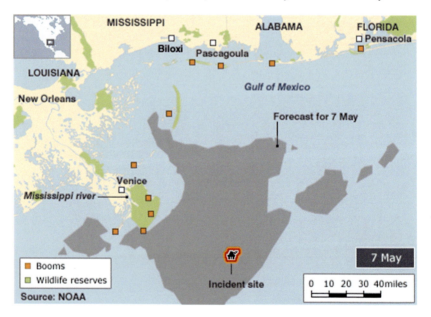

FIGURE 6.13 Location of protective booms on May 7 (NOAA/BBC).

Oil Spill Response and Cleanup Methods 173

FIGURE 6.14 Deployment of a boom in a U configuration (USCG Atlantic Area).

in the deployment of booms to prevent oil reaching onshore and islands as shown in Figure 6.15. Figure 6.16 shows a warehouse in Louisiana for the storage of oil-absorbent booms. Figures 6.17 and 6.18 show a center in Alabama for washing and cleaning the booms for re-deployment as the vice president visited the center. A vessel places a containment boom in Louisiana's Barataria Bay in a photo taken

FIGURE 6.15 An oil containment boom deployed by the US Navy surrounds New Harbor Island in Louisiana to control environmental damage from BP oil spill (photo by US Navy, Wikimedia Commons).

FIGURE 6.16 Oil-absorbent boom in a warehouse in Amelia, La., May 13 (photo by Patrick Kelley of the US Coast Guard, USGC/BBC).

by the US Coast Guard (Figure 6.19). In Figures 6.20–6.24 the cleaning of booms at a center in Louisiana as well as the deployment and training of crews of shrimp boats are shown. Throughout the response to the GOM oil spill, the Coast Guard was actively involved in the cleanup efforts (Figure 6.25). Hundreds of miles of boom were deployed along the Gulf coast, but politicians were asking for more of the highly visible barriers (Deep Water, 2011). By June 25, the total length of boom deployed to prevent oil reaching the coast was 2.8 million feet and about 4.2 million

FIGURE 6.17 Workers pressure wash an inflatable boom at the Theodore Staging Area in Alabama, July 22, 2010 (official White House photo by David Lienemann).

Oil Spill Response and Cleanup Methods 175

FIGURE 6.18 Vice President Joe Biden speaks to workers and the press at the Theodore Staging Area in Alabama, July 22, 2010 (official White House photo by David Lienemann).

feet of absorbent boom. In the case of the BP oil spill, booms were partly successful, although rough seas washed oil over them.

6.4 IN-SITU BURNING

According to the Office of Response and Restoration of NOAA (https://response.restoration.noaa.gov/) in-situ burning is a technique that involves the controlled burning of oil at the location of the spill. If conducted properly, it significantly reduces the

FIGURE 6.19 Placing boom along Louisiana coast was used as a barrier to prevent oil reaching the coast (Deep Water 2011/National Commission Report to President Jan 2011).

FIGURE 6.20 In this photo workers at a decontamination site in Venice, La., clean an oil-containment boom on Tuesday May 4, 2010 (USCG photo).

FIGURE 6.21 Shrimp boats tow fire-resistant oil-containment boom as the crews conduct in-situ burn training off the coast of Venice, La., on May 3, 2010 (USCG photo by Patrick Kelley).

Oil Spill Response and Cleanup Methods 177

FIGURE 6.22 Workers at a decontamination site carry an oil-containment boom that was cleaned in Venice, Louisiana, May 4, 2010. The boom was transferred to a staging area where it was put back into service aboard one of many boats fighting to mitigate the BP oil spill (US Coast Guard photo by Patrick Kelley).

FIGURE 6.23 An oil containment boom at Pensacola, Fla., by Naval Air Station on May 4 (photo by US Navy/Patrick Nichols).

FIGURE 6.24 Members of Naval Air Station Pensacola Pollution Response unit set an oil-containment boom May 4, 2010 at Sherman Cove, Pensacola, Fla., to protect sensitive areas from the Deepwater Horizon oil spill (US Navy photo by Patrick Nichols).

amount of oil on the water and minimizes the adverse effect of the oil on the environment, although burning itself could contribute to air pollution. In-situ burning usually is conducted after the oil is concentrated by fire-resistant booms. Figures 6.26–6.29 show fields of fire on the Gulf of Mexico where officials were hoping to minimize the amount of oil reaching the shores. As shown in these figures the US Coast

FIGURE 6.25 The US Coast Guard supported cleanup efforts after the DWH oil spill (USCG photo by Patrick Nichols).

Oil Spill Response and Cleanup Methods

FIGURE 6.26 Clouds of smoke billow up from controlled burns taking place in the Gulf of Mexico May 19, 2010 (Wikinews, May 25).

Guard, working with BP, local residents, and federal agencies, conducted the burn to help prevent the spread of oil following the April 20 explosion of Deepwater Horizon.

In-situ burning of thick and fresh oil can be initiated very quickly by igniting the oil with ignition devices. In-situ burning can remove oil from the water surface very efficiently and very rapidly. Removal efficiencies for thick slicks can easily exceed 90%. Removal rates of 2000 m^3/hour can be achieved with a fire area of only about

FIGURE 6.27 Smoke billows over a controlled oil fire in the Gulf of Mexico off Venice, La., May 5, 2010 (photo by US Navy by Justin E. Stumber).

FIGURE 6.28 Controlled burn of oil from the DWH/BP/GOM oil spill (US Coast Guard photo).

FIGURE 6.29 In-situ burning was employed extensively during the Deepwater Horizon/BP oil spill in 2010 (US Coast Guard from NOAA Response).

10,000 m^2 or a circle of about 100 m in diameter (EMSA, 2013). In Figure 6.30 controlled burning is shown in the Gulf of Mexico. Finally the flaring of gas from the drillship Discoverer Enterprise is shown in Figure 6.31 and the ship is shown in Figure 6.32. Gas and oil are brought to the surface by a tube placed inside the damaged pipe as a method to collect some oil and gas from the well.

Oil Spill Response and Cleanup Methods 181

FIGURE 6.30 Controlled burn of oil spilled in the Deepwater Horizon disaster, Gulf of Mexico, May 6, 2010. The burning oil was contained by a length of boom (EPA/Britannica).

6.5 DISPERSANTS

Dispersants (a kind of surfactant) are used with a solvent by spraying through aircraft (Figure 6.33) or by boats (Figure 6.34). Dispersants can slow down the movement of oil slicks to the coastal areas and wetlands but they cannot remove the oil from the sea. They in fact relocate the oil deeper in the water column.

FIGURE 6.31 Gas from the damaged Deepwater Horizon wellhead is burned by the drillship *Discoverer Enterprise* May 16, 2010, in a process known as flaring. Gas and oil from the wellhead were being brought to the surface via a tube that was placed inside the damaged pipe (US Coast Guard photo by Patrick Kelley, USGS, June/2010).

FIGURE 6.32 The drillship Discoverer Enterprise was used in efforts to control the spreading crude, setting parts of the oil slick on fire to prevent it from reaching shore. This picture is from May 23, 2010 (Photo by US Coast Guard).

Dispersants when applied to the water surface are droplets of 0.4–1 mm in diameter. Solvents (usually glycol or light petroleum distillates) help to deliver surfactants to the oil/water interface. Surfactants are detergent-like molecules with two tails. One tail is a lipophilic (synonymous with oleophilic) molecule which likes oil, fat, and hydrocarbons; the other tail of the dispersant is hydrophilic which likes water

FIGURE 6.33 Airplane releasing oil dispersant (source: US Coast Guard, credit: Stephen B. Lehmann/Southern Louisiana University, https://www2.southeastern.edu/orgs/oilspill/cleanup.html).

FIGURE 6.34 Spraying dispersants by boat (ITOPF).

molecules. Dispersants can reduce interfacial tension between oil and water and can break oil molecules into smaller particles and cause them to disperse into the water column. Broken oil molecules with a size of 10–50 μm (micrometers) disperse into the water column while smaller broken oil molecules of less than 1 μm form a thin sheen layer of oil on the water surface. Types of surfactant and solvents with their availability are given in Table 6.1 (EMSA, 2010). Only second- and third-generation dispersants are now available to be applied to oil spills. The second generation are also called Type 1, while the third generation may have Types 2 and 3 according to the UK system. Applications of these types of dispersants are given in Table 6.2. In the Gulf of Mexico, UK-manufactured dispersants, like Corexit 9500/9527, were used which were a mixture of solvents and surfactants (NAP, 2014). Table 6.2 also shows under what oil properties the use of dispersants may be successful. For example for very highly viscous oil with kinematic viscosity of greater than 10,000 cSt, the use of dispersants is not possible. Table 6.3 shows treatment rates for each type of dispersant and under different conditions with suggested spraying methods (ESMA 2010).

The application of a dispersant depends on the sea conditions as well. For example if the temperature is high and light components of oil can be easily vaporized, dispersants should be used in the first few hours or few days, right after the oil spill occurs, because once light components of oil are vaporized, the remaining oil becomes more viscous and the effectiveness of dispersants reduces as shown in Figure 6.35 based on series of tests conducted by SINTEF (Trondheim, Norway). Figure 6.36 shows the unsuccessful application of a dispersant to high-viscosity oil in the field. Table 6.4 shows a combination of effects of oil viscosity and the sea state on the feasibility of the use of dispersants. The typical concentration of oil versus time after treatment with dispersants is given

TABLE 6.1
Type of Surfactants and Solvents Used in Dispersants

Generation	UK Type	Surfactants	Solvents	Current Availability
First-generation dispersants		(i) Nonylphenol ethoxylates or tall oil ethoxylates	Kerosene extract (KEX) kerosene	No longer used as oil spill dispersants
"Conventional" or hydrocarbon-base" or "second-generation dispersants"	UK Type 1 "conventional" or "hydrocarbon-base" dispersants	(i) Fatty acid esters (ii) Ethoxylated fatty acid esters	Light petroleum distillate: Odorless or de-aromatized kerosene Low aromatics (less than 3% wt.) kerosene CAS No 64742-47-8 EC No 265-149-8	Available
"Concentrate" or "third-generation dispersants"	UK Type 2 "water-dilutable concentrate dispersants"	(i) Fatty acid esters or sorbitan esters such as Span™ series CAS No. 1338-43-8 (ii) Ethoxylated fatty acid esters (PEG esters) or ethoxylated sorbitan esters such as Tween™ series CAS No 103991-30-6 (iii) Sodium di-iso-octyl sulphosuccinate EC NO. 209-406-4 CAS No 577-11-7	Glycol ethers such as: Ethylene glycol Dipropylene glycol 2-butoxyethanol (Butyl Cellosolve™) CAS No. 111-76-2 Di-propylene glycol monomethyl ether CAS No 34590-94-8 EC No 252-104-2	Available
	UK Type 3 "concentrate" dispersants		Light petroleum distillates: Hydrotreated light distillates CAS No 64742-47-8 EC NO. 265-149-8	Available

Source: EMSA (2010).

TABLE 6.2
Type of Surfactants and Solvents Used in Dispersants

Dispersant				Spilled oil			
				Viscosity of Spilled Oil (in cSt at Sea Temperature)			
				Possibility for Dispersion			
Generation	Description	UK Type	Sprayed from	Less than 500 — Generally Easy	500–5000 — Generally Possible	5000–10,000 — Sometimes Possible	Greater than 10,000 — Generally Impossible
Second	"Conventional" or "hydrocarbon-base"	UK Type 1	Ships, boats, onshore	Dispersant effective at 30% volume treatment rate	Dispersant effective at 30–50% volume treatment rate	Dispersant possibly effective at 100% volume treatment rate	Ineffective
Third	"Concentrate" or "water- dilutable concentrate"	UK Type 2	Ships and boats	Dispersant effective at 50–100% volume treatment rate	Ineffective	Ineffective	Ineffective
	"Concentrate"	UK Type 3	Aircraft, ships, and boats	Dispersant effective at 5% volume treatment rate	Dispersant effective at 5–10% volume treatment rate	Dispersant effective at 10–15% volume treatment rate	Ineffective

Source: EMSA (2010).

TABLE 6.3

Type of Spraying Method and Treatment Rate for Dispersants

Description and Generation	UK Type	Sprayed from	Recommended Treatment Rate	Comments	Current Availability
First-generation dispersants		Ships, boats, onshore	High treatment rate 30–50% dispersant as volume of spilled oil or 1 part dispersant to 2 to 3 parts oil	Industrial detergents with solvents that are too toxic to be used as dispersants	No longer used as oil soil dispersants
"Conventional" or "hydrocarbon-base" or "second-generation dispersants"	UK Type 1 "conventional" or "hydrocarbon-base" dispersant	Ships, boats, onshore	High treatment rate 30–50% dispersant as volume of spilled oil or 1 part dispersant to 2 to 3 parts oil	Low toxicity High treatment rate	Available
"Concentrate" or "third-generation dispersants"	UK Type 2 "water-dilutable concentrate dispersant	Ships and boats	10S 10% solution of dispersant in seawater to 2 to 3 parts oil Equivalent to 1 part dispersant to 20 to 30 parts oil	Low toxicity High treatment rate when diluted with water and can only be sprayed from ships and boats in this way	Available
	UK Type 3 "concentrate" dispersant	Aircraft, ships, and boats	Low treatment rate 3 to 5% dispersant as volume of spilled oil or 1 part dispersant to 20 to 30 parts oil	Low toxicity Low treatment rate Used undiluted (or neat)	Available

Source: EMSA (2010).

Oil Spill Response and Cleanup Methods

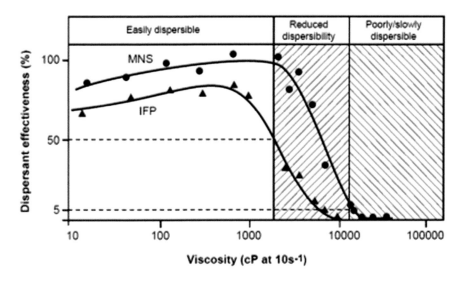

FIGURE 6.35 Effect of oil viscosity on dispersant effectiveness (EMSA, 2010).

in Table 6.5. The effectiveness of use of dispersants versus type of accident is given in Table 6.6. These guidelines are not very quantitative, and the assessment to apply a certain method relies on expert judgment (Zeinstra-Helfrich, 2016).

Some other general guidelines for the application of dispersants are given below (EMSA, 2010):

- Dispersants should not be used in very shallow water.
- Dispersants should not be used in an oil spill that is directly above shellfish beds.

FIGURE 6.36 High oil viscosity has resulted in unsuccessful dispersant application, noted by the typical white plume of dispersant around the oil (source: ITOPF).

TABLE 6.4

Feasibility of Chemical Dispersant Application Based on Sea State and Oil Viscosity

Sea State/Oil Viscos.	0–1	2–3	>3
<500 cSt	Possible on limited spill with mixing energy	Easy	Easy
500–5000 cSt	None	Possible	Easy
5000–10,000 cSt	None	Uncertain	Possible
>10,000 cSt	None	None	Uncertain
>>10,000 cSt	None	None	None

Source: EMSA (2016).

- Dispersants should not be used on spilled oil that is directly over corals, sea grass, and fish.
- The use of dispersants close to industrial water intakes which are normally protected by fixed booms is not advisable.

For the case of the BP oil spill, the NOAA's Office of Response and Restoration conducted aerial dispersant observer training to prepare flight crews for improving the effectiveness of dispersant sorties (NOAA, 2020).

The fate of dispersants used for oil dispersion is also important from an environmental point of view due to the toxicity of chemicals used. Water and air temperatures, type of dispersant, wind speed, and wave action have a good impact on the fate of dispersants as they eventually biodegrade. In the case of Deepwater Horizon or similar cases, the oil source is deep below the water surface at the sea floor where the temperature is lower and the oil is more viscous, and hence the effectiveness of surfactants is lower. In response to the DWH oil spill, about 800,000 gallons of dispersants were used near the seafloor (Kujawinski et al.,

TABLE 6.5

Concentration of Oil in Water after Dispersant Application

Time after Dispersant Application	Oil Concentration in the Upper Water Column in ppm
Just after treatment	10
2 days after treatment	1
1 week after treatment	0.5
1 month after treatment	0.2
3 months after treatment	Background level

Source: EMSA (2010).

Oil Spill Response and Cleanup Methods 189

TABLE 6.6
Effectiveness of Dispersant Treatment versus Types of Accident and Spilled Oil

Incident Involving	Spilled Oil	Dispersant Use Effective or Appropriate?
Fishing vessel	Marine diesel oil	No
	Marine gas oil	
Small cargo ship	Medium fuel oil	Yes
Medium cargo ship	Medium fuel oil	Yes
Product tanker	Medium / Heavy Fuel Oil	Yes
Product tanker	Gasoline cargo	No
Product tanker	Jet fuel cargo	No
Product tanker	Diesel cargo	No
	Vegetable oil	No
Product tanker	HFO for power use	No
Large cargo ship	Heavy fuel oil	Possibly
Oil tanker	Heavy fuel oil	Possibly
Oil tanker	Condensate	Probably no
Oil tanker	Crude oil cargo	Yes – for some time

Source: EMSA (2010).

2011). Before this incident, dispersants were never used in a location so deep (5000 ft) in water. A key ingredient of the dispersants is dioctyl sodium sulfosuccinate (DOSS) which according to the EPA degrades rapidly under conditions similar to the DWH oil spill. White et al. (2014) studied the fate of dispersants used in DWH and concluded that DOSS can remain on GOM beaches for nearly four years and up to the level of 260 ng/g.

Generally a choice to use dispersants is not clear, and it is a compromise with other options considering cost and protecting other sources from oil pollution, and only an expert can decide considering all factors discussed above.

6.6 SKIMMING AND MECHANICAL REMOVAL METHODS

Skimming generally represents a variety of processes that use mechanical tools to physically remove and collect floating oil from the water surface. The equipment used for such processes is usually referred to as skimmers in which a conveyor belt is used to remove and collect the oil before reaching sensitive areas. Another example of such tools is weir skimmers which gather oil into underground storage tanks mainly through the gravity forces. Other tools include suction or vacuum machines and the use of shovels to remove and collect oil. The selection of a mechanical method versus a chemical method (the use of dispersants or in-situ burning) mainly depends on the thickness of the oil and sea conditions as shown in Figure 6.37. Generally, as shown in this figure, mechanical methods are effective

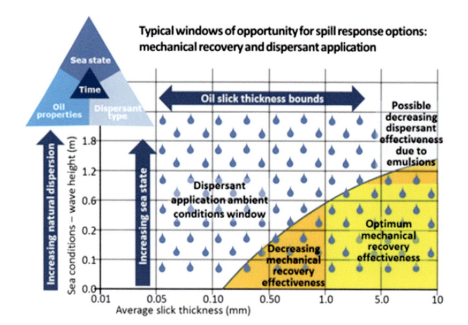

FIGURE 6.37 Applicability of oil mechanical and chemical oil spill response as a function of sea conditions and slick thickness (source: www.amsa.gov.au, Zeinstra-Helfrich, 2016).

when oil thickness is between 1 and 10 mm and the sea conditions are calm with waves of less than 1 m height.

In Figure 6.38, oil is being skimmed from the sea surface by a "vessel of opportunity." Sometimes, two boats will tow a collection boom, allowing oil to concentrate within the booms, where it is then picked up by a skimmer as shown in Figure 6.39. A giant vessel called "A Whale" is designed to vacuum up oily water, separate the oil, and return the water to the sea as shown in Figure 6.40. With a length of 3 1/2 football fields and towering 10 stories high, "A Whale" is designed to collect up to 500,000 barrels of oily water a day through 12 vents on either side of its bow. The ship was used in the Gulf of Mexico to assist the cleanup of the BP oil spill. A self-propelled weir skimmer for use in ports and ashore waters is shown in Figure 6.41. The bow doors open to enhance the swath and allow entry of floating oil, and recovered oil is pumped to an internal storage tank.

Figures 6.42 and 6.43 show scenes where workers contracted by BP are working in a massive cleanup operation for the DWH oil spill on Louisiana beaches during the months of May and June 2010 in the midst of the disaster. Figure 6.44 shows the removal of oil on shores using vacuums by cleanup workers. Five weeks after the accident, the operating company BP announced its plan for the building of tent cities and even floating camps to house thousands of cleanup workers so they could spend more time at work (ABC, May 28). Figures 6.45 and 6.46 show two other methods of removing oil from the sea and beaches using shovels and machinery.

Oil Spill Response and Cleanup Methods 191

FIGURE 6.38 A boat, the "vessel of opportunity," skims oil spilled after the Deepwater Horizon/BP well blowout in the Gulf of Mexico in April 2010 (NOAA, April 2010).

FIGURE 6.39 Curtain boom employed in a V configuration by two towing vessels with a separate skimming vessel (source: ITOPF).

FIGURE 6.40 The "A Whale" skimmer ship, the world's largest oil skimming vessel, skims oil, seen in the foreground, in the Gulf of Mexico July 3, 2010 (source: UPI/A.J. Sisco).

Finally, another method which was used in the case of the BP oil spill to prevent oil reaching shores was the use of sand berms. Berms form a barrier usually by sand in order to stop oil approaching coasts. However, the oil should be removed mechanically by other methods such as the use of shovels as shown in Figure 6.45. Any oil that does wash up is likely to be in the form of tar balls (Figure 6.10) which are easier to handle but if not removed could stay for very long time (BBC, May 18).

FIGURE 6.41 A self-propelled weir skimmer for use in ports and ashore waters (source: ITOPF).

Oil Spill Response and Cleanup Methods

FIGURE 6.42 Deepwater Horizon oil spill: beach cleanup workers contracted by BP cleaning up oil on a beach in Port Fourchon, Louisiana, May 23, 2010. (PO3 Patrick Kelley/US Department of Defense, Commons Wikimedia).

FIGURE 6.43 A cleanup crew member rakes in and collects oil waste in Grand Isle, La., as part of the DWH/BP oil spill response on May 27, 2010 (US Coast Guard photo by Ann Marie Gorden).

FIGURE 6.44 Workers attempt to vacuum the oil from the Gulf of Mexico (source: Flickr/Kris Krug).

FIGURE 6.45 Cleanup teams used shovels and their hands to gather affected soil and ocean debris along oil-impacted beaches north of Santa Barbara, California, May 21, 2015 (US Coast Guard/NOAA).

Oil Spill Response and Cleanup Methods

FIGURE 6.46 Heavy machinery was brought in to remove oil from a beach in Puerto Rico in 2007 (NOAA).

6.7 SUMMARY

Once a major oil spill occurs near residential and business areas, such as for the case of Deepwater Horizon, personnel have to be deployed quickly to protect the shoreline and wildlife. Vessels including skimmers and barges in addition to aircraft and remotely controlled vehicles are needed for responding to the site of the accident. Other physically effective tools to prevent the oil spill reaching coastlines are containment booms and sorbent booms, and the use of dispersants as well as the creation of staging areas to protect sensitive shorelines. However, before taking any actions, it is important to simulate the behavior of the spill under natural conditions as shown in Figure 6.47 and in the next two chapters. Figure 6.47 shows a time trajectory of oil removal by three methods and the amount remaining for a time period of 76 days according to the model studied by You and Leyffer (2011) and Zhong and You (2011).

The projected behavior of an oil spill along with the location of spill and weather conditions will determine what should be the best and effective cleanup method considering the cost and availability of personnel and equipment needed. For the case of Deepwater Horizon almost three-quarters of the oil spilled in the Gulf of Mexico was cleaned up or broken down by natural forces based on a US government report (BBC, August 5). As shown in Figure 6.48 and according to the NOAA predictions of what happened to the 4.9 million barrels of oil leaked into the sea in August 2010, about 17% of the oil was removed by inserting a tube, 25% evaporated, 5% burned off at the surface, 3% skimmed from the surface, 8% chemically dispersed, 16% naturally dispersed, and 26% remained residually or washed onshore. The response to the BP oil spill was unique in the length of containment booms (13 million feet), the vast amount

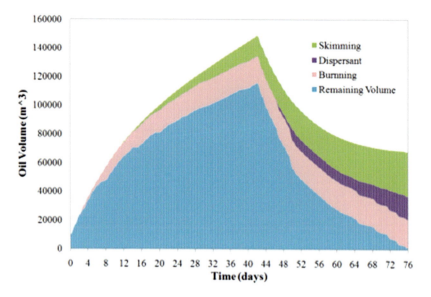

FIGURE 6.47 Time trajectories of the oil volumes removed by three methods and remaining on the sea surface when the response time span is 76 days (You and Leyffer, 2011).

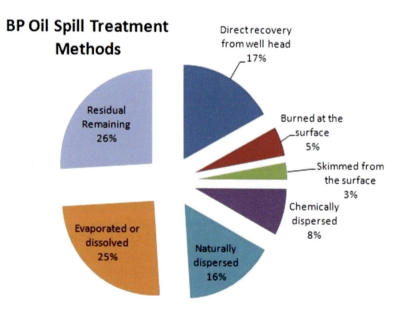

FIGURE 6.48 Distribution of BP oil spill treatment methods (source: Kerr, 2010, NOAA Model).

Oil Spill Response and Cleanup Methods

of equipment provided from all over the world, thousands of vessels involved in recovery and burning, hundreds of aircraft, and the amount of dispersants used, as some estimates put it at more than 1.9 million gallons. The spill covered an area of 57,000 square miles and 1100 miles of shoreline were polluted (Britannica).

REFERENCES

Barrios, D.A.J. 2016. *Numerical simulation of oil spills.* Master Thesis Universidad Politecnica de Madrid, Spain. http://oa.upm.es/47049/1/TFM_Donaldo_Augusto_Juvinao_Barrios.pdf

CBS News, 2010, Public Offers Government Ideas to stop oil flow., May 6. www.cbsnews.com/news/public-offers-govt-ideas-to-stop-oil-flow/

EMSA. 2016. *Overview of national dispersant testing and approval policies in the EU.* European Maritime Safety Agency. March. file:///C:/Users/User/Downloads/Overview%20of%20dispersant%20tests%20in%20Europe%20(1).pdf

EMSA. 2013. *European Maritime safety agency.* As adopted by EMSA's Administrative Board at its 37th Meeting held in Lisbon, Portugal, on November 13–14. p. 38. file:///C:/Users/User/Downloads/action-plan-for-response-to-marine-pollution-from-oil-and-gas-installations%20(1).pdf

EMSA. 2010. *Manual on the applicability of oil spill dispersants - version 2.* European Maritime Safety Agency 2010. www.emsa.europa.eu file:///C:/Users/User/Downloads/dispersant_manual_web.pdf

ITOPF. 2018. *Use of dispersants to treat oil spills.* Technical Information Paper 4. London. https://www.itopf.org/fileadmin/data/Documents/TIPS%20TAPS/TIP_4_Use_of_Dispersants_to_Treat_Oil_Spills.pdf, file:///C:/Users/User/Downloads/TIP_2_Fate_of_Marine_Oil_Spills%20(1).pdf

Kerr, R.A. 2010. Gulf oil spill: A lot of oil on the loose, not so much to be found. *Science*, 329(5993), pp. 734–735. Published 8.13.10.

Kujawinski, E.B. et al. 2011. Fate of dispersants associated with the deepwater horizon oil spill. *Environmental Science & Technology*, 45(4), pp. 1298–1306.

NAP. 2014. *Responding to oil spills in the U.S.* Arctic Marine Environment, The National Academies Press, Washington, DC.

NCAR, National Center for Atmospheric Research, University Corporation for Atmospheric Research. https://ncar.ucar.edu/

NOAA. 2020. *Spill containment methods.* July 14. https://response.restoration.noaa.gov/oil-and-chemical-spills/oil-spills/spill-containment-methods.html

NRC. 2003. *National Research Council (US) Committee on oil in the sea: Inputs, fates, and effects.* National Academies Press (US), Washington, DC. https://www.ncbi.nlm.nih.gov/books/NBK220700/

White, H.K., S.L. Lyons, S.J. Harrison, et al. 2014. Long-term persistence of dispersants following the deepwater horizon oil spill. *Environmental Science & Technology Letters*, 1(7), pp. 295–299.

You, F. and S. Leyffer. 2011. Mixed-integer dynamic optimization for oil spill response planning with integration of a dynamic oil weathering model. *AIChE Journal*, 57, pp. 3555–3564.

Zeinstra-Helfrich, M. 2016. *Oil slick fate in 3D, predicting the influence of (natural and chemical) dispersion on oil slick fate.* PhD Thesis, Wageningen University, Netherlands. https://edepot.wur.nl/389993

Zhong, Z. and F. You* 2011. Oil spill response planning with consideration of physicochemical evolution of the oil slick: A multiobjective optimization approach. *Computers & Chemical Engineering*, 35, pp. 1614–1630.

7 Simple Models to Predict Rate of Oil Spill Disappearance

NOMENCLATURE

A	Oil spill surface area
A_o	Initial oil spill surface area
API	API gravity (= $141.5/SG - 131.5$) (dimensionless)
C^{vap}_{int}	Oil molar concentration in air at spill interface (gmol/L)
C_s	Oil solubility in water (gmol/L)
d	Liquid density at 20°C and 1 atm, g/cm^3
F_V	Volume fraction of oil disappeared, dimensionless
$F(P)$	Probability density function for property P
K^{dis}	Mass transfer coefficient for the rate of dissolution (m/s)
K^{vap}	Vaporization mass transfer coefficient (m/s)
M	Molecular weight of oil (g/gmol)
m	Mass of oil
n	Number of moles of oil sample (gmol)
n_o	Initial number of moles of oil sample (gmol)
P^{sat}	Vapor pressure of oil at sea surface temperature
P_c	Critical pressure
Q^{vap}	Parameter defined in Equation 5.5 (1/day or 1/s)
R	Gas constant (8.314 Pa m^3/mol-K)
r	Rate of oil disappearance (1/day or 1/s)
SG	Specific gravity at 15.5°C
S_w	Salt concentration in seawater, wt%
T	Temperature (K)
T_c	Critical temperature (K)
V	Oil spill volume (m^3)
V_o	Initial oil spill volume (m^3) or cm^3
U	Wind speed (m/s)
x_i	Mole fraction of pseudocomponent i in oil spill (dimensionless)
y	Oil spill thickness (m)
Z^{sat}	Liquid saturated liquid compressibility factor

GREEK LETTERS

ρ_m Oil molar density (gmol/cm^3)
ρ_{liq} Liquid absolute density (g/cm^3)
Δt Time step, hour or s
ω Acentric factor, dimensionless

SUPERSCRIPTS

vap   Vaporization
dis   Dissolution

SUBSCRIPTS

i   Pseudocomponent properties
r   Reduced parameter
v   A value based on volume
o   Initial values (variable at $t=0$)

ACRONYMS

BP   British Petroleum
GOR   Gas-to-oil ratio
GOM   Gulf of Mexico
NOAA   National Oceanic and Atmospheric Administration

7.1 INTRODUCTION

Oil spills can be monitored by GIS which is a computer system capable of analyzing and storing geographic data. An oil spill can also be detected by SAR which is an active sensor that captures microwaves from the target objects. The use of SAR permits the detection of oil pollution on the sea surface day and night and in most weather conditions (Automatic Oil Spill Detection 1999). SAR image classification can distinguish between three different pollution levels for an oil spill: the spill area in the center, the surrounding high-pollution area, and the outer layers of the low-pollution area (Mansor et al. 2006).

The fate of an oil spill depends on the type and composition of the oil and the weather conditions such as the temperature and wind speed. Light oils are more toxic than heavier oils; however, heavy oils stay longer on the water surface. For example, oil spills in the Persian Gulf stayed for a very long time after they occurred. The Gulf War oil spill was one of the worst oil spills in history, resulting from the actions taken during the Gulf War in January 23, 1991, and it did considerable damage to the environment in the Persian Gulf. The exact size of the spill remains unknown; however, the estimate of 6–8 million barrels (about 1 million m^3) is mostly referenced.

Simple Models to Predict Rate of Disappearance 201

The Gulf of Mexico (GOM) oil spill, also known as the Deepwater Horizon oil spill or the BP oil spill, was a result of an accident on April 20, 2010, in the Macondo well located 52 miles southeast of the Louisiana port of Venice and at a depth of 5,000 ft. This was the worst oil spill accident in US history and released nearly 5 million barrels of oil until it was sealed on July 15, 2010, as reported by NOAA (2010). NOAA is a federal agency focused on the condition of the oceans and the atmosphere (www.noaa.gov) and was the main government agency monitoring the situation during the crisis (NOAA, 2010). Other estimates put the number for the amount of oil leaked at $4.4 \times 10^6 \pm 20\%$ barrels and at a rate of $5.6 \times 10^4 \pm 21\%$ bbl/day (Crone and Tolstoy, 2010).

As discussed in Chapter 6, when no action is taken, the fate of a spill of crude oil or refined products in the marine environment is determined by spreading, evaporation, dissolution, dispersion, emulsification, sedimentation, and biodegradation and photo-oxidation processes (Kuiper and Van den Brink, 1987). A schematic of such processes that demonstrates the fate of an oil spill is shown in Figure 7.1 (Riazi, 2010). The simulation and prediction of the fate of an oil spill is crucial for the planning of the response to the accident and choosing the right cleanup operation. There is a number of different methods which might be used to deal with oil floating on the sea. Some of these methods are burning, mechanical recovery with skimmers, oil removal with the use of absorbents, use of chemical agents, to make it into a gel and then skim it from the surface, to sink it to the bottom, and to emulsify or disperse it (Breuel 1981). Perhaps the cheapest and easiest way to remove an oil spill from seawater is to vaporize and disperse it. Light petroleum fractions such as gasoline or kerosene can be completely vaporized with time. Crude oils which consist of some heavy compounds may not be completely vaporized, as such components tend to disperse, degrade, or sediment rather than to vaporize or dissolve in water. Prediction of the fate of an oil spill is important in selecting an appropriate cleanup method.

Once an oil spill occurs, oil rapidly spreads over a large area and breaks up in long narrow slicks with the same direction as the wind. Spreading of oil causes an increase in the evaporation of light components from the oil. Dissolution is a physical process that causes some low molecular weight components to dissolve in water and is strongly affected by the physiochemical conditions and is much slower than evaporation. Dissolution is much slower than evaporation, and the rate of oil dissolution is less than 0.1% of the rate of evaporation. But it is important to know the amount of dissolution because of the toxicity of the dissolved oil components. While air temperature affects the rate of evaporation, water temperature affects the rate of oil dissolution. The components which dissolve in water are in contact with water and the oil molecules that evaporate are in contact with air in the outer layer of slick.

Degradation is the decomposition of oil through oxidation, especially photo-oxidation under the ultraviolet waves of sun. Some heteroatoms in the oil such as sulfur slow down the rate of oxidation, while vanadium enhances the degradation process. Components produced from the degradation of oil have higher solubility in water with a higher degree of toxicity than the original components. Sedimentation is the process of the deposition of oil at the bottom of sea. Some 10–30% of oil may be adsorbed on suspended materials and finally deposited at the bottom of the sea. In

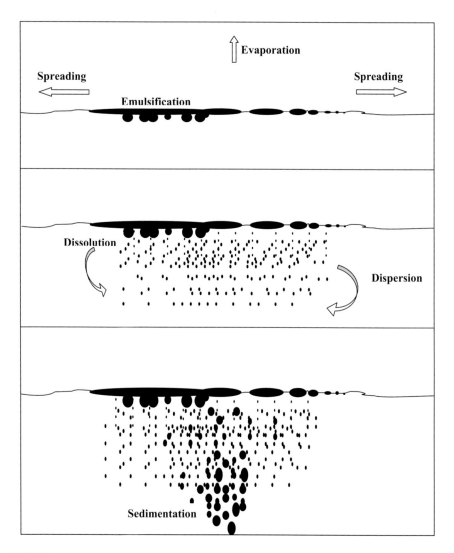

FIGURE 7.1 Behavior of an oil spill on seawater surface

addition, some particles of oil heavier than water (with specific gravity greater than or equal to unity) may deposit. Finally some heavy components in the oil may go through another physical process called aggregation. Oil may aggregate as a petroleum lump or tar balls in water. The basis of aggregation is the presence of heavy molecules such as wax or asphaltene (see Figure 2.1). Oil aggregates have a size of 1 mm–10 cm and look gray. All the above processes are dynamic and depend on time as well as the rate of spreading of oil on the seawater surface. Spreading and motion of oil on the sea happen due to gravitational forces and are controlled by oil viscosity and surface tension. Only ten minutes after a spill of 1 ton of oil, the oil can disperse over a radius of 50 m, forming a slick 10 mm thick. The slick gets thinner (less than

Simple Models to Predict Rate of Disappearance 203

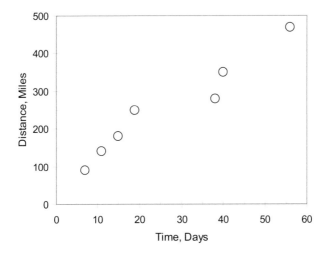

FIGURE 7.2 The rate of spreading and flowing for the *Exxon* oil spill.

1 mm) as the oil continues to spread, covering an area of up to 12 km² (Patin, 1999). For the Alaska oil spill, the rate of spreading was estimated based on published data (Galt et al. 1991, Leber 1989). The length of the spill was increased from 50 miles to 500 miles after two months as shown in Figure 7.2 (Riazi and Roomi, 2008).

An ideal model for the fate of an oil spill should simulate mathematically all of the above dynamic processes versus time. State-of-the-art models include some but not all of these processes in a single model. Field experiments designed to test these models usually focus on one process as studied and reviewed by Spaulding (1988), Psaltaki and Markatos (2005), Villoria et al. (1991), Riazi and Edalat (1996), Nasr and Smith (2006) and Riazi and Al-Enezi (1999), and Riazi (2000, 2016).

Heavy compounds and residues tend to disperse into water or to sink into the bottom of the sea. The rate at which a hydrocarbon dissolves in water is generally lower than the rate of evaporation under the same conditions (Wheeler 1987, Green and Trett 1989, ASCE 1996). It is widely considered that, after volatility, the most significant property of oil components, from the point of view of their behavior in aquatic environments, is their solubility in water (Green and Trett 1989). It has been shown by Riazi and Al-Enezi (1999) and Riazi and Roomi (2008) that the rate of oil dissolution in water is small in comparison with the rate of oil evaporation and the rate of oil dissolution in calculation of the overall rate of oil disappearance may be insignificant. For this reason many numerical models developed for oil spill trajectory do not consider the dissolution process. But the knowledge of dissolved hydrocarbons concentration in water is important from a toxicological and environmental point of view. Aromatic hydrocarbons, especially mono-aromatics such as benzenes, are the most toxic compounds, and their amount in water determines the degree of toxicity in water. The physical process of dissolution is well understood, but the description in the case of oil spills is complicated, due to the complex oil composition with hundreds of components and the necessity of describing the dissolution of a single component with component-specific parameters. The most soluble oil components are usually the most toxic. Even low concentrations of these toxic compounds could lead to serious effects on biological systems and could be toxic to fish and wildlife.

Riazi and Al-Enezi (1999) proposed mathematical relations for the rates of oil evaporation and dissolution by introducing two temperature-dependent mass transfer coefficients, one for evaporation and one for dissolution with variable slick thickness. They also produced some experimental data on the rate of disappearance of some petroleum products and a crude oil floating on water at different conditions. They also developed a more complete model to consider the rate of sedimentation as well as the multi-component nature of crude oils, but they did not consider dissolution of various hydrocarbon compounds. The models presented in this chapter are a compilation of the physical models developed in the past two decades and may be used for a quick prediction of the fate of crude oil spills on the sea surface under various conditions before any cleanup operation.

Although the term oil spill generally refers to a crude oil spill, it can also be used in more general terms for the cases of petroleum products (fuels and non-fuels), liquid chemicals and petrochemicals, and liquefied natural gas (LNG) (such as liquefied ethane, propane, and butane) as well as other liquid fuels such as biofuels and coal liquids. Methods presented in this chapter for the rate of evaporation may also be applied to cases of onshore oil spills in which one can determine how much of the oil is vaporized and how much remains on the ground versus time.

7.2 MODELING SCHEME

Consider an oil spill floating on the seawater surface with an initial volume of V_o and initial surface area of A_o in which the subscript "$_o$" indicates an initial value at time zero. We assume the oil is a mixture of N components/pseudocomponents each specified by "i." Then at any time t after the occurrence of the oil spill, its volume (V), area (A), and the thickness (y) are given as follows:

$$V = \sum_{i=1}^{N} V_i, \qquad A = \sum_{i=1}^{N} A_i, \qquad y = \frac{V}{A} \frac{1}{2} \tag{7.1}$$

For each component the volume fraction disappeared after time t is defined as $F_{Vi} = (V_{oi} - V_i) / V_{oi}$. In the mathematical model we choose a time step of Δt and a semi-analytical approach is used to formulate the model. During each time step it is assumed that the slick thickness is constant. The volume fraction of each component x_{vi} is calculated from V_i and V as $x_{vi} = V_i/V$. The volume of each component V_i can be converted to mass (m_i) through its density ρ_i ($m_i = V_i \times \rho_i$). The total mass of oil at any time is calculated from the mass of individual components: $m = \sum m_i$. A schematic of the model is shown in Figure 7.3 where the oil is assumed as a mixture of N components. Component 1 is the lightest, while N is the heaviest component in the oil. Components with specific gravity SG_i ($= \rho_i/\rho_{\text{water}}$) greater than unity sink to the bottom of the sea immediately. It is assumed that all components heavier than

Simple Models to Predict Rate of Disappearance

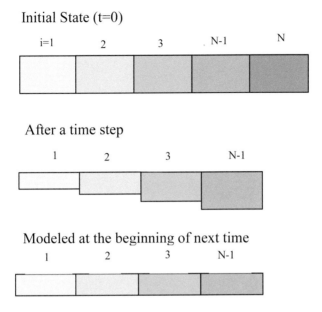

FIGURE 7.3 Modeling scheme for an oil spill.

water sink within the first time step. Based on the remaining components, a new slick thickness can be determined (Riazi, 2000, Riazi and Al-Enezi, 1999). The total volume fraction of oil evaporated at time t is calculated from

$$F_V = \sum_{i=1}^{N} x_{vi} F_{Vi} \tag{7.2}$$

The rate of mass transfer through vaporization can be expressed by the following mathematical relation (Riazi and Al-Enezi, 1999):

$$J_i^{vap} = -\frac{1}{A_i}\frac{dn_i}{dt} = K_i^{vap}\left(C_{i,s}^{vap,air} - C_i^{vap,air}\right) \tag{7.3}$$

where J_i^{vap} is the mass transfer flux due to vaporization, n_i is the number of moles of component i in the spill vaporized, K_i^{vap} is the mass transfer coefficient for vaporization, $C_{i,s}^{vap}$ is the concentration of component i in the air at the water surface, and C_i^{vap} is the concentration of i in the air far from the surface. $C_{i,s}^{vap}$ can be calculated from the vapor pressure of component i while C_i^{vap} can be considered zero. This equation can be converted into the following form in terms of the fraction of volume change of the oil spill, F_{vi}^{vap}, for a time step of Δt during which the surface area, thickness, and temperature are constant:

$$F_{Vi}^{vap} = 1 - \exp(-Q_i^{vap}\Delta t) \tag{7.4}$$

where Q_i^{vap} is a dimensionless parameter related to the mass transfer coefficient K_i^{vap} in the air which depends on the diffusion coefficient of hydrocarbons in the air and wind speed (Riazi and Al-Enezi, 1999).

$$Q_i^{vap} = \frac{K_i^{vap} Z_{liq,i}^{sat}}{y_i} \tag{7.5}$$

$$K_i^{vap} = \frac{1.5 \times 10^{-5} T U^{0.8}}{M_i^2} \tag{7.6}$$

$$Z_{liq,i}^{sat} = \frac{P_i^{vap} M_i}{\rho_{liq,i}^{sat} RT} \tag{7.7}$$

In these relations, U is the wind speed, M_i is the molecular weight, $Z_{liq,i}$ is the saturated liquid compressibility factor for component "i" at temperature T, and R is the gas constant. $\rho_{liq,i}^{sat}$ is the saturated liquid density of component "i" which can be calculated from the Rackett equation as discussed in Chapter 2 (Equation 2.61).

$$V^{sat} = \left(\frac{RT_c}{P_c}\right) Z_{RA}^{n} \tag{7.8}$$

$$n = 1.0 + (1.0 - T_r)^{2/7}$$

where T_r is the reduced temperature and Z_{RA} is the Rackett parameter which can be estimated from the specific gravity as given in Chapter 2 (Equation 2.62).

In Equation 7.7, P^{vap}_i is the vapor pressure of component "i" at temperature T which may be estimated from the bulk properties of pseudocomponent "i" using the Lee–Kesler method from Chapter 2 (Equation 2.67):

$$\ln P_r^{vap} = 5.92714 - 6.09648 T_r^{-1} - 1.28862 \ln T_r + 0.169347 T_r^6 \tag{7.9}$$
$$+ \omega(15.2518 - 15.6875 T_r^{-1} - 13.4721 \ln T_r + 0.43566 T_r^6)$$

In a similar fashion the rate of dissolution for each component "i" can be formulated as described by Riazi and Al-Enezi (1999):

$$J_i^{dis} = K_i^{dis} (C_{si} - C_{wi}) \tag{7.10}$$

where J_i^{dis} is the molar flux of dissolution for component i. K_i^{dis} is the mass transfer coefficient due to dissolution which is related to the diffusion coefficient of component i in water. C_{si} is the molar solubility of the oil in water and C_{wi} is the concentration of component i in water which can be considered as zero as discussed by Riazi and Edalat (1996). Then the volume fraction removed by the dissolution of oil (F_{Vi}^{dis}) during each time step Δt is calculated from the following relation:

Simple Models to Predict Rate of Disappearance

$$F_{Vi}^{dis} = 1 - \exp(-Q_i^{dis} \Delta t) \quad \text{where} \quad Q_i^{dis} = \frac{K_i^{dis} C_{si}}{y_i \rho_{mi}} \tag{7.11}$$

In the above equation ρ_{mi} is the liquid molar density and C_{si} is the molar solubility of component "i" in water at the specific water temperature and K_i^{dis} is the mass transfer coefficient defined in Equation 7.10. Based on the diffusion coefficients discussed in Chapter 2, Riazi and Edalat (1996) derived the following relation for calculation of K_i^{dis}:

$$K_i^{dis} = \frac{4.18 \times 10^{-9} T^{0.67}}{V_{Ai}^{0.4} A_i^{0.1}} \tag{7.12}$$

in which K_i^{dis} is in m/s, T is in K, V_{Ai} is the molar volume of component i at its normal boiling point in m³/gmol which can be estimated from Equation 7.8. A_i is the surface area of component i in m².

Once the volume fraction of each component dissolved is calculated from Equations 7.4 and 7.11, the mass of each component vaporized (m_i^{vap}) or dissolved (m_i^{dis}) can be calculated through the following relations:

$$m_i^{vap} = \rho_i F_{vi}^{vap} Voi \tag{7.13}$$

$$m_i^{dis} = \rho_i F_{Vi}^{dis} V_{oi} \tag{7.14}$$

The total mass of oil vaporized or dissolved can be calculated from the following summations:

$$m^{vap} = \sum m_i^{vap} \tag{7.15}$$

$$m^{dis} = \sum_{i=1}^{N} m_i^{dis} \tag{7.16}$$

Similarly the mass of oil deposited as sediment can be calculated from the sum of masses of components with $SG_i > 1$. The total mass of spill disappeared after time t is then calculated through the sum of masses disappeared due to evaporation, dissolution, and sedimentation for all components. The calculation procedure of the parameters required by this model is discussed in the next section.

7.3 CALCULATION PROCEDURE

In this model an oil spill is considered as a mixture of different real and pseudo-components. These pseudocomponents are treated separately, and those components which are heavier than water with specific gravities greater than one would immediately sink into water. Although the sedimentation of oil into water is a time-dependent process, in this model it is considered as an instant process for the sake of mathematical simplicity. So at the beginning of the calculations, if the density of the component "i" is greater than water density at the same temperature, then the component sinks into the water and it does not enter the subsequent calculations for the rates of evaporation and dissolution. After each time step, similar calculations are repeated beginning with the last value of slick thickness calculated from the previous time. A summary of the calculation procedure is shown in Figure 7.4. To demonstrate the calculation procedure, the composition and characteristics of a crude oil and four petroleum products as reported by Riazi and Al-Enezi (1999) are given in Tables 7.1 and 7.2.

As shown in Table 7.1, the composition of a crude oil is normally represented through the mole fraction of pure light components from ethane to pentane, lumped hexanes (C_6), and heptane-plus fraction (C_{7+}), where C_6 is a narrow-cut hydrocarbon mixture while C_{7+} is a wide hydrocarbon mixture containing hydrocarbons from heptane and heavier. In addition to the composition, the molecular weight and specific gravity of heptane-plus fractions (M_{7+}, SG_{7+}) are also known from laboratory data. The probability density function (PDF) discussed in Chapter 2 (Equation 2.34) can be used to split the C_{7+} fraction.

$$ F(P^*) = \frac{B^2}{A} P^{*B-1} \exp\left(-\frac{B}{A} P^{*B}\right) \tag{7.17} $$

where P is a property such as absolute boiling point (T_b), molecular weight (M), or specific gravity (SG). In this equation, P^* is a dimensionless form of parameter P ($P^* = (P - P_o)/P_o$) in which P_o is the lightest component in the C_{7+} fraction. Using this PDF, the C_{7+} can be represented by a number of pseudocomponents with known boiling point (T_b) and specific gravity (SG) as given in Table 7.3 for the crude oil in Table 7.1. These two parameters are sufficient to estimate all necessary physical properties using the methods presented in Chapter 2.

In the use of Equation 7.11 for the calculation of amount of dissolution, the solubility of oil in water C_s is needed. Riazi and Al-Enezi (1999) measured the solubility of the above crude oil and four petroleum products in water at temperatures from 20 to 50°C and salt concentration from 0 to 4%. Generally temperature increase causes an increase of the oil solubility in water, while the presence of salt causes a decrease in the solubility. Based on these data they suggested the following relation to estimate the solubility of a hydrocarbon component or a pseudocomponent from its molecular weight at any given temperature and salt concentration.

$$ C_{si} = \exp[(4.6 - 0.0036 M_i) + (0.1 - 0.0018 M_i)S_w - 4250 / T] \tag{7.18} $$

in which C_{si} is in mol/L, T is the absolute temperature of water in K, and S_w is the salt concentration in water in weight%. As shown in the above correlation, heavier components have lower solubility.

The amount of oil dissolved in water can be calculated from Equation 7.14; however, each pseudocomponent is a mixture of hydrocarbons from the groups of paraffins (P), napthenes (N), and aromatics (A). Aromatic hydrocarbons are more toxic

Simple Models to Predict Rate of Disappearance

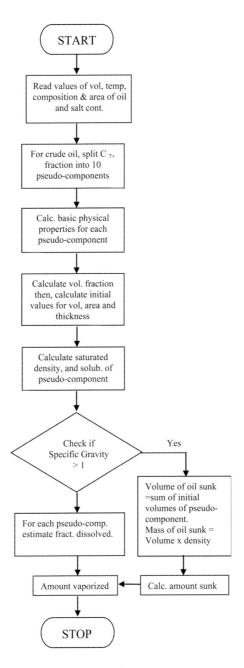

FIGURE 7.4 Summary of calculation procedure.

TABLE 7.1
Composition and General Characteristics of a Kuwaiti Crude Oil for Export

Component	Mole%
Ethane (C_2)	0.2
Propane (C_3)	2.0
Iso-butane (iC_4)	2.4
n-Butane (nC_4)	4.2
Iso-pentane (iC_5)	2.4
n-Pentane (nC_5)	4.1
Hexanes (C_6)	5.3
Heptane-plus (C_{7+})	79.4
Total	100.0

Source: Riazi and Al-Enezi (1999).

For the crude: molecular weight = 210, API = 31, specific gravity = 0.8708, sulfur weight % = 2.4%.

For C_{7+} fraction: molar weight = 266.6, specific gravity = 0.8910, weight average boiling point = 603.2 K.

TABLE 7.2
General Characteristics of Kuwaiti Petroleum Products for Export

Property	Sample			
	Naphtha	Kerosene	Diesel Fuel	Gas Oil
ASTM-D86 distillation (°C) (%)				
0	65.6	147.2	223.9	274.4
5	81.1	162.8	257.8	351. 1
10	83.3	167.8	268.9	371.1
30	91.7	180.0	293.3	396.7
50	100.6	195.6	310.0	434.4
70	112.8	214.4	332.2	456.1
90	128.9	235.6	364.4	492.2
95	135.6	243.3	379.4	513.3
100	152.8	257.2	388.9	530.0
Specific gravity (15.5°C)	0.7188	0.7898	0.8589	0.9259
Density (20°C)	0.7136	0.7867	0.8568	0.9225
Refractive index (20°C)	1.4049	1.4410	1.4804	1.5262
Molecular weight	101.3	155.7	243.7	394.3

Source: Riazi and Al-Enezi (1999).

Simple Models to Predict Rate of Disappearance

TABLE 7.3

Characterization Parameters for Pseudocomponents of the C_{7+} Fraction of Kuwaiti Crude Oil

Pseudocomponent	MoN%	Weight%	Molecular weight	Specific gravity	T_b (°C)
C_{7+} (1)	16.0	8.2	116.1	0.734	128
C_{7+} (2)	12.0	7.6	142.2	0.773	123
C_{7+} (3)	11.2	9.0	173.9	0.807	221
C_{7+} (4)	9.2	11.8	236.5	0.855	300
C_{7+} (5)	8.9	9.4	331.2	0.907	393
C_{7+} (6)	6.9	18.0	456.3	0.937	482
C_{7+} (7)	3.6	18.2	593.1	0.980	560
C_{7+} (8)	3.6	11.6	727.4	1.049	632
Total C_{7+} fraction	79.4	93.8	266.1	0.891	330

Source: Riazi and Al-Enezi (1999).

than paraffinic and naphthenic hydrocarbons. In addition, mono-aromatics (MA) are more toxic than polyaromatics (PA). Studies also found that PA hydrocarbons can cause lung as well as other forms of cancer. Benzene compounds are very stable and difficult to break down. For this reason, in order to determine the degree of toxicity of dissolved oil in water, knowledge of the amount aromatics (A) and mono-aromatics (MA) dissolved is important. The PNA composition can be calculated from Equations 2.37–2.45, and Equations 2.46–2.48 can be used to estimate fractions of MA, PA, and aromatics for oil fractions with molecular weight less than 250 as given below:

$$x_{MA} = -62.8245 + 59.90816R_i - 0.0248335m \tag{7.19}$$

$$x_{PA} = 11.88175 - 11.2213R_i + 0.023745m \tag{7.20}$$

$$x_A = x_{MA} + x_{PA} \tag{7.21}$$

Application of these methods to both crude oil and its products is demonstrated in the next section in comparison to laboratory data.

7.4 LABORATORY EXPERIMENTS AND MODEL PREDICTIONS

Riazi and Al-Enezi conducted some experiments with the crude oil shown in Table 7.1 and the petroleum products in Table 7.2. One experiment was conducted at 42°C, with an average wind speed of 5 m/s. The initial volume of the oil spill was 500 mL (cm³) and the initial area was 3116 cm². Water salinity was 3% by weight which was taken from Maseela Beach off the coast of Kuwait. The initial area was 3116 cm² and

TABLE 7.4

Comparison of Model Predictions with Experimental Values for the Volume Fraction Removed for Kuwait Export Crude at 42°C after 174 h

	Experimental	Calculated
Initial volume of oil spill (cm³)	500	500
Initial oil spill diameter (cm)	63	63
Initial slick thickness (mm)	1.6	1.6
Volume of oil vaporized (cm³)	218	237
Volume of oil sunk (cm³)	56	49
Volume of oil dissolved (cm³)	0.76	0.9
Volume of oil remaining (cm³)	225.24	213.1
Volume % disappeared ($100\,F_v$)	55.0	57.4
Diameter of oil spill (cm)	60.5	60.2
Slick thickness (mm)	0.78	0.75

Source: Riazi and Al-Enezi (1999).

the volume of oil on the water surface after 174 hours was 0.76 cm³ as given in Table 7.4. Predicted data from the above model are also given in this table. Details of the experimental procedure and measurements with a step-by-step calculation method are given by Riazi and Al-Enezi (1999).

The model prediction for the volume fraction removed (F_v) versus time for the crude oil at 22 and 42°C is shown in Figure 7.5. Experimental data are also shown in the same figure. The progression of slick thickness versus time for the crude oil at 42°C is shown in Figure 7.6. The predicted slick thickness was reduced from 1.6 to 0.75 mm while measured values showed a reduction from 1.6 to 0.78 mm, confirming the validity of calculations for the specific crude oil sample.

The rate of disappearance (F_v) for the diesel oil (Table 5.2) at 22°C is shown in Figure 7.7. The crude oil average molecular weight is 210 while the molecular weight of the diesel is 244, and as such the crude contains more volatile components. For this reason, at 22°C the rate of disappearance of the crude oil is higher than the rate of diesel. These data and calculations show a relatively good agreement between model prediction and laboratory data for a fixed initial volume of oil spill floating on the water surface.

Based on the crude oil and the petroleum products with the specifications given in Tables 7.1 and 7.2 and with the use of Equations 7.14–7.16, the concentration of mono-aromatics (MA) and polyaromatics (PA) dissolved in water can be calculated. For the case of the crude oil sample (Table 7.1 and Table 7.3), the amount of oil remaining on water surface versus time at 42°C is shown in Figure 7.8. As shown in Table 7.3, the last two pseudocomponents have specific gravity greater than or equal to unity ($SG \geq 1$), and 30% by weight of the initial oil may settle or disperse in

Simple Models to Predict Rate of Disappearance 213

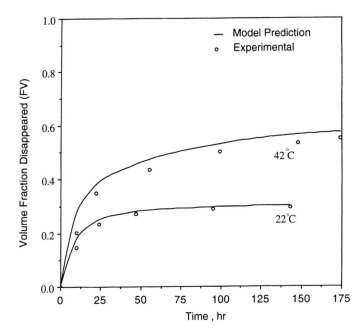

FIGURE 7.5 Rate of disappearance of Kuwaiti crude (Table 7.4) oil spill at 22 and 42°C (Riazi and Al-Enezi, 1999).

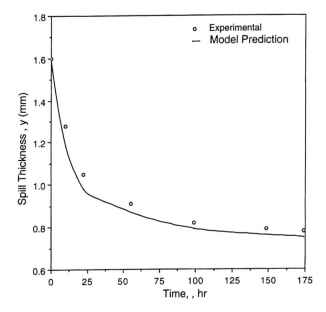

FIGURE 7.6 Rate of change of slick thickness for Kuwaiti crude (Table 7.4) oil spill 42°C (Riazi and Al-Enezi, 1999).

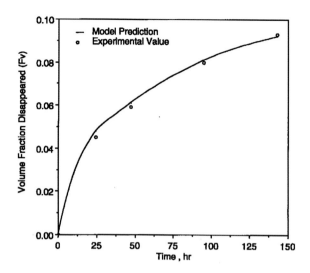

FIGURE 7.7 Rate of disappearance for Kuwaiti diesel spill at 22°C (Riazi and Al-Enezi, 1999).

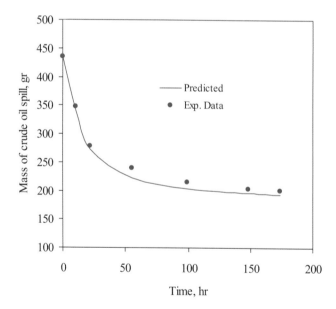

FIGURE 7.8 Model prediction for the mass of crude oil spill remaining on the water surface at 42°C. Predicted (———), Exp. Data (•).

water (corresponding to 25% of initial oil by volume). The initial oil mass was 435 gr; therefore the amount of oil sedimentation was approximately 51 gr. The volume of oil deposited was measured as 56 mL while the predicted value was 49 mL (Table 7.4). This difference is partly due to the characterization of oil and number of the pseudocomponents as presented in Table 7.3.

Simple Models to Predict Rate of Disappearance

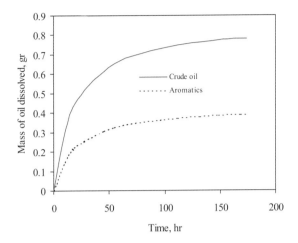

FIGURE 7.9 Model prediction for the mass of crude oil and aromatic hydrocarbons dissolved in water at 42°C for a spill area of 0.32 m². Crude oil (———), aromatics (----------).

The amount of oil dissolved and the amount of aromatics dissolved in water versus time are shown in Figure 7.9. Generally the amount of aromatics in oil increases with the molecular weight of oil. The amount of mono-aromatics (MA) dissolved with respect to total aromatic compounds at 42°C is shown in Figure 7.10. The rate of dissolution of aromatics and MA compounds at 22°C is shown in Figure 7.11. As shown in these figures, the amount of dissolution increases significantly with temperature. The amount of dissolution of mono-aromatics is important from a toxicological point of view as they are the most toxic compounds in a crude oil sample.

As mentioned earlier (Equation 7.18), the presence of salt in water can reduce the solubility of oil. This is demonstrated in Figure 7.12 for the case of naphtha oil in

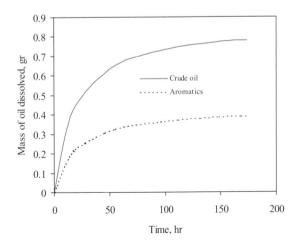

FIGURE 7.10 Model prediction for the rate of oil dissolution at 42°C. Total compounds (———), total aromatics (--------), mono-aromatics (- - - -).

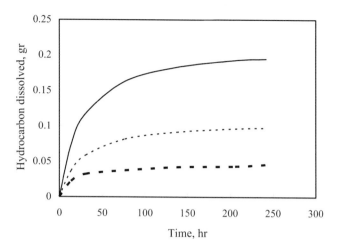

FIGURE 7.11 Model prediction for the rate of oil crude dissolution at 20°C. Total compounds (———), total aromatics (--------), mono-aromatics (- - - -).

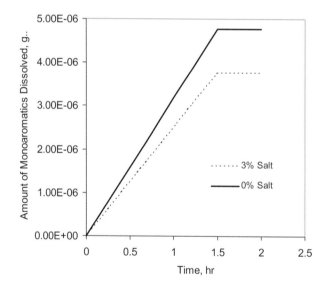

FIGURE 7.12 Rate of dissolution of mono-aromatics in water at 32°C from a naphtha oil spill with an area of 0.32 m². 0% salt (———), 3% salt (--------).

Table 7.2. Similar results are shown in Figure 7.13 for the amount of kerosene sample dissolved at 32°C. As shown in these figures, the amount of oil dissolved in water is very small. The reason for this is that light products, such as naphtha, in the oil spill can be completely vaporized within few hours, and consequently the amount of dissolved compounds remains constant after the complete disappearance of the oil spill due to the rapid vaporization process.

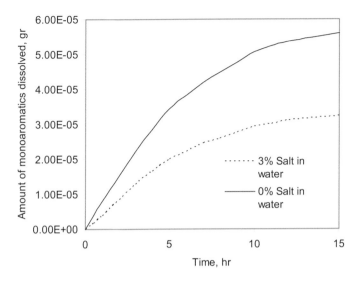

FIGURE 7.13 Rate of dissolution of mono-aromatics in water at 32°C from a kerosine oil spill with an area of 0.32 m². 0% salt (———), 3% salt (--------).

7.5 APPLICATION OF THE MODEL TO THE CASE OF CONTINUOUS FLOW OF OIL

During the accident at Macondo well in the Gulf of Mexico in April 2010, oil leaked at a rate of 57,000 barrels per day on a continuous basis until the well was closed on July 15, 2010. This was the worst oil spill accident in US history and released 4.9 million barrels of oil until it was sealed after 86 days on July 15, 2010 (about 57,000 bbl/day or 377 m³/h) as reported by NOAA. Information about this oil spill and the properties of the oil released and meteorological data were presented in Chapter 4. A large portion of this oil (40°API) was vaporized during a period in which temperature varied between 20 and 34°C.

The model presented in this chapter is applied to this case in which oil is flowing for a period of nearly 3 months. In this modeling approach, temperature and wind speed are considered variable on a daily basis, and a time step of 12 hours (half a day) is chosen for the calculation of the rate of evaporation of hydrocarbons from the oil spill floating on the sea surface coming from a source on a continuous basis. The most readily available data for the crude oil produced from the Macondo Well are its density or API gravity. However, Reddy et al. (2011) reported a detailed GC analysis of oil and gas produced from the well at 5,000 ft below the water surface. They found that the fluid flowing out of the Macondo well had a gas-to-oil ratio (GOR) of 1,600 standard cubic feet per barrel. The compositions of the oil, gas, and reservoir fluid are given in Table 4.3, while the general characteristics of the oil are given in Table 4.4 in Chapter 4. The oil API gravity is 40 (equivalent to a density of 0.85 g/cm³) as reported by Reddy and US government organizations (NOAA, 2010). These data were used with the density function given by Equation 5.17 to generate molar

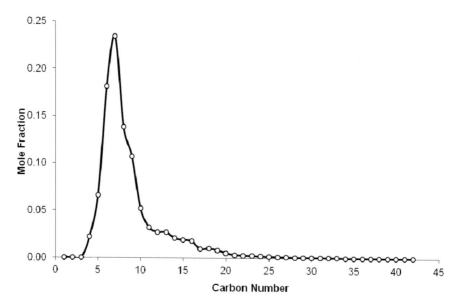

FIGURE 7.14 Molar distribution of hydrocarbons in the oil.

distribution. Based on the normalization of the data, the molar distribution of the crude is presented in Figure 7.14. The crude oil in the GOM spill was divided into 19 pseudocomponents with those heavier than water considered sedimented and the remaining components involved in the vaporization process according their vapor pressure and mass transfer coefficient in the air as described by Riazi (2016).

The variation of temperature and wind speed in Venice, La., from May to July 2010 was recorded, and it is given in Chapter 4 (Table 4.6). The spread between low and high temperatures for this period is shown in Figure 7.15 as reported by Riazi (2016). For the computational calculations, these temperature variations have been correlated to time (day) as given in Table 7.5. The variation of wind speed for the same period is shown in Figure 7.16. Based on the area of spill and its volume as reported by NOAA, the initial slick thickness can be estimated.

An Excel sheet was specifically developed to perform the calculations for this problem. The calculation results for the volume of oil vaporized versus time are presented in Figure 7.18. The days are counted starting from April 20 when the oil flow began. The volume of oil was calculated from the sum of volume of all pseudocomponents. In Figure 7.18, the horizontal parts indicate no vaporization due to zero wind speed for those specific days as explained above.

In the proposed model the initial oil thickness (y) is needed. This was calculated from Equation 7.1 and based on the spill area from Figure 7.17 on April 26. The initial oil thickness was calculated as 0.03 mm. According to Table 5.3, the last two components have $SG \geq 1$ and therefore account for 24.9% of the total oil volume removed before vaporization began. So the model prediction for the amount of sedimentation (which may be considered as the amount dispersed) is about 25% of the initial volume. According to the data released by NOAA (2010), about 24% total oil

Simple Models to Predict Rate of Disappearance

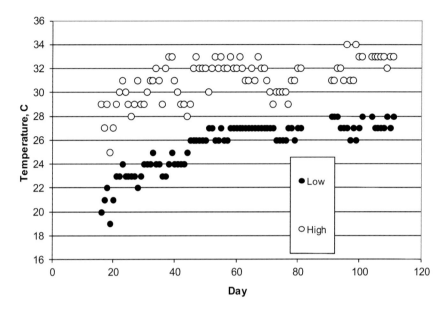

FIGURE 7.15 The variation of temperature versus time in days (April 20, 2010, Day = 0).

TABLE 7.5
Low and High Temperature Based on Data from Table 4.6 (Chapter 4) and Figure 7.15

T (°C) = a + b (Day/100) + c (Day/100)^2

	a	b	c	d	AD, °C
Low temperature	15.73	36.55	−39.96	14.85	0.56
High temperature	21.47	47.71	−73.71	36.59	1.00

---

was dispersed (16% naturally and 8% by chemicals) while 25% vaporized and 26% remained at the sea or on shore as shown in Figure 7.19.

In the calculations using the above models it was assumed that the oil is either vaporized or sedimented, and Figure 7.18 shows that about 50% of the total oil released into the sea was vaporized while 25% was sedimented and 25% remained in the sea under natural conditions. According to NOAA data (Figure 7.19), 50% of the oil was removed or burned or dispersed. If this amount (half of 4.9 million barrels of oil released) is considered in these calculations (after 25% of the initial oil was dispersed or sunk) then the model estimate would be 25% vaporized, which agrees with the NOAA estimate. It is not expected to achieve identical predictions, as in the actual case, the rate of oil released varied in the early and final stages of the accident. In the last few weeks, as attempts to seal the pipe were underway, the amount of oil

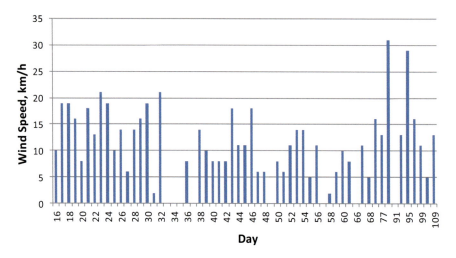

FIGURE 7.16 The variation of wind speed versus time in days (April 20, 2010, Day = 0).

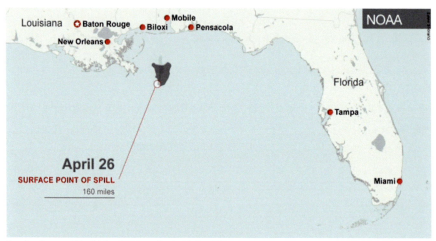

Based on the area of this slick the oil thickness is calculated as 0.03 mm.

FIGURE 7.17 Position and area of the oil spill after four days (source: NOAA [2010]). Based on the area of this slick the oil thickness is calculated as 0.03 mm.

released was reduced gradually until July 15, when oil flow was completely stopped. In addition, other factors such as dissolution, the timing of removal, or burning also affect the calculation results which are not considered here.

7.6 SUMMARY

In this chapter some simple and semi-analytical models were presented for a quick and rough estimate of the amount of oil vaporized, dissolved, or deposited as sediment versus time when an oil spill occurs on the water surface. In addition the model

Simple Models to Predict Rate of Disappearance

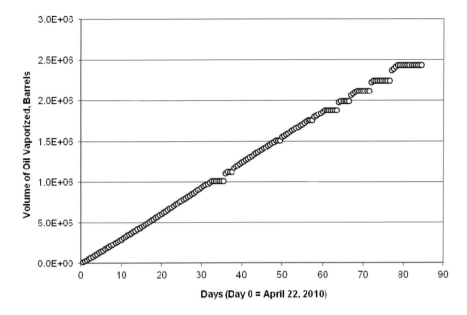

FIGURE 7.18 Model prediction for the amount of oil vaporized versus time.

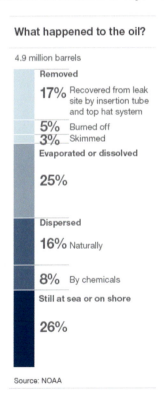

FIGURE 7.19 Fate of the GOM oil spill according to NOAA estimates (2010).

calculates the oil spill area, and its volume and thickness at any time after the oil is released on the water surface. Data for crude oils and petroleum products from the Persian Gulf area have been used to evaluate the model prediction when a fixed amount of oil is released on the sea surface. The amount of toxic compounds dissolved in water were also calculated under different conditions and salt concentration. The presence of salt reduces the amount of oil dissolution in water. As seen from the results, the level of dissolved hydrocarbons is small because of the low solubility of hydrocarbons in water, but knowledge of this concentration is important from an environmental point of view, especially for the amount of aromatics and mono-aromatics in water because of the toxic nature of these compounds.

The model was also extended for the cases in which oil flows continuously such as the case of the Gulf of Mexico (BP) oil spill, and the results were compared with predictions from some commercial models. In this model a time step of 12 hours (half a day) was used for the calculations. The model requires minimum data, usually available for crude oils, along with air and water temperatures, wind and water speeds, and initial volume and area of the spill. According to the model prediction for the case of the GOM oil spill, after removing half of the spill by dispersion or combustion, about 25% of the initial oil is vaporized and 18% persists in comparison to 25% vaporized and 26% persisting according to the NOAA model predictions.

REFERENCES

ASCE Task Force. 1996. State-of-the-art review of modeling transport and fate of oil spills. *Journal of Hydraulic Engineering*, 122 (11), November, pp. 594–609.

Automatic Oil Spill Detection, ENVISYS. 1999. *The Norwegian Computing Center (NR)*. http://www2.nr.no/envisys/emergency_management/marine_oil_spill.htm

Breuel, A. 1981. How to dispose of oil and hazardous chemical spill derbis. *Pollution Technology Review*, 87, pp. 37–71.

Crone, T.J. and M. Tolstoy. 2010. Magnitude of the 2010 gulf of mexico oil leak. *Science*, 330(6004), p. 634. doi:10.1126/Science.1195840.

Galt, J.A., W.J. Lehr and D.L. Payton. 1991. Fate and transport of Exxon Valdez oil spill. *Environmental Science & Technology*, 25(2), p. 202.

Green, J. and M.W. Trett. 1989. *The fate and effects of oil in freshwater*. Elsevier, London.

Kuiper, J. and W.J. Van den Brink. 1987. *Fate and effects of oil in marine ecosystems*. Martinus Nijhoff Publishers, Boston.

Leber, P.A. 1989. *Environmental chemistry: A case study of the Exxon Valdez oil spill of 1989*. Franklin and Marshall College, Lancaster, PA. http://wulfenite.fandm.edu/exxon-valdez.htm

Mansor, S.B., H. Assilzadeh and H.M. Ibrahim. 2006. Oil spill detection and monitoring from satellite image. *GIS Development 2006*. http://www.gisdevelopment.net/application/miscellaneous/misc027.htm

Nasr, P. and E. Smith. 2006. Simulation of oil spills near environmentally sensitive areas in Egyptian coastal waters. *Water and Environment Journal*, 20(1), pp. 11–18.

NOAA – National Oceanic and Atmospheric Administration, US Department of Commerce. Washington, DC. http://response.restoration.noaa.gov/oil-and-chemical-spills (accessed on June 3, 2010).

Patin, S. 1999. *Environmental impact of the offshore oil and gas industry*, translated by E. Cascio. http://www.offshore-environment.com/oil.html

Psaltaki, M.G. and N.C. Markatos. 2005. Modeling the behavior of an oil spill in the marine environment. *IASME Transactions*, 2(9), pp. 1656–1664.

Reddy, C.M., J.S. Arey, J.S. Seewald, S.P. Sylva, K.L. Lemkau, R.K. Nelson, C.A. Carmichael, C.P. McIntyre, J. Fenwick, G.T. Ventura, B.A.S. Van Mooy and R. Camiili. 2011. *Composition and fate of gas and oil released to the water column during the Deepwater Horizon oil spill*, edited by J.M. Hayes. PNAS: Proceedings of National Academy of Sciences of the United States

Riazi, M.R. 2000. An improved mathematical model for the fate of a crude oil spill. *Belgian Journal of Operations Research, Statistics and Computer Science (JORBEL)*, 40(3–4), pp. 123–135.

Riazi, M.R. 2010. Accidental oil spills and control. In *Environmentally conscious fossil energy production*, edited by Myer Kutz. John Wiley and Sons, New York, NY, January.

Riazi, M.R. 2016. Modeling and predicting the rate of hydrocarbon vaporization from oil spills with continuous oil flow. *International Journal of Oil, Gas, and Coal Technology (IJOGCT)*, 11(1), January, pp. 93–105.

Riazi, M.R. and G. Alenzi. 1999. A mathematical model for the rate of oil spill disappearance from seawater for Kuwaiti crude and its products. *Chemical Engineering Journal*, 73, pp. 161–172.

Riazi, M.R., Edalat, M. 1996. "Prediction of the Rate of Oil Removal from Seawater by Evaporation and Dissolution," *Journal of Petroleum Science and Engineering*, 16, pp. 291–300.

Riazi, M.R. and Y.A. Roomi. 2008. A model to predict rate of dissolution of toxic compounds into seawater from an oil spill. *International Journal of Toxicology*, 27(5), pp. 379–386.

Spaulding, M.L. 1988. A state-of-art review of oil trajectory and fate modeling. *Oil and Chemical Pollution*, 4, pp. 39–55.

Villoria, C.M., A.E. Anselmi, S.A. Intevep and F.R. Garcia. 1991. An oil spill fate model. *SPE23371*, pp. 445–454.

Wheeler, R.B. 1987. *The fate of petroleum in the Marine environment*. Production Research Company Special Report, Exxon, New Jersey.

8 Advanced Oil Spill Modeling and Simulation Techniques

Konstantinos Kotzakoulakis[a,b]
and Simon C. George[c]

[a]CICESE Physical Oceanography,
Ensenada, Baja California, Mexico

[b]SINTEF Ocean AS, Brattørkaia 17c, 7010
Trondheim, Norway (current position)

[c]Department of Earth and Environmental Sciences
and Macquarie University Marine Research Centre

8.1 INTRODUCTION

Oil spill modeling is an essential tool in the effort to minimize the environmental and financial impacts of accidental oil releases. It has multiple applications, spanning from oil spill preparation and response to forensic investigation and scientific research. In more detail, oil spill modeling is used in:

- Statistical studies in order to identify the likelihood of an area to be hit by an oil spill, the potential risk, and the corresponding impact that an oil spill will have on local species, ecosystems, and financial activities. This information is used by the relevant authorities to plan and prepare for potential oil spill events.
- Oil spill forecasting during an accident, in order to guide the responding authorities to decide the most effective counter measures for mitigating the effects of the oil spill.
- Forensic investigation for the identification of the origin of an unknown oil spill, by running reversed time simulations to determine the past trajectory of the oil spill.
- Scientific studies that investigate the combined effects of different environmental processes and conditions on the circulation and fate of the oil spill, especially when actual field measurements are scarce for the condition under investigation.

During the past decade, there has been intense research activity in the field of oil spill modeling. Most of the environmental processes that affect the fate of oil spills have been revisited, and new processes have been investigated. Research activity was intensified after the 2010 Deepwater Horizon disaster in order to address the lack of knowledge on the fate of deep-water oil releases and the effectiveness of response measures such as subsea dispersant injection. The Deepwater Horizon oil spill was the largest and best-documented oil spill event to date (e.g. Camilli et al., 2010; Hazen et al., 2010), and provided a large amount of valuable field data for modelers to use in the development of the next-generation oil spill models.

Although historically most oil spills happened at the sea surface, and previous generations of oil spill models treated them as two-dimensional sea-surface-only simulations, there are strong reasons that support the importance of subsurface oil releases and the need for 3D oil spill modeling in general. Subsea oil spills usually happen on the seafloor when the well head isolation mechanism (the blowout preventer) fails to stop a sudden high-pressure gas-induced surge into the wellbore. The source of the oil in these leaks is the underground oil reservoir, and the amount of oil that can potentially leak is much greater in comparison to a typical surface oil spill from an oil tanker. Additionally, the technical challenges that must be overcome to successfully stop the oil leak are greater, and usually these leaks last longer. The importance of this type of simulation is becoming even greater due to the current trend in the oil industry to explore further offshore in deeper waters, because the swallower discoveries are gradually getting depleted and deeper unexplored regions can reveal large petroleum reservoirs.

Finally, better understandings of certain environmental processes, such as the wave-induced entrainment of oil in the water column, indicate the need for 3D oil spill modeling even for the cases of sea-surface oil spills. Consequently, the contemporary modeling approach is the use of 3D modeling for both surface and subsurface oil spills.

In the following sections, modern techniques to simulate different types of oil leaks will be presented as well as some of the applications of oil spill modeling mentioned earlier.

8.2 NEAR-FIELD AND FAR-FIELD OIL SPILL MODELING

When simulating subsea oil leaks, models are divided into two distinct spatio-temporal domains (Figure 8.1). The rapid changes the pressurized oil undergoes when it escapes into the marine environment and the evolution of the buoyant oil plume at the initial stages belong to the near-field model domain. The spatio-temporal extent of this domain is usually in the order of hours and a few hundred meters, and it ends when the oil plume has lost its initial momentum and has become neutrally buoyant. The far-field domain starts from this point on, when the oil is now advected by water currents, and if it reaches the surface, by the wind and the waves as well. The rapid transformations of oil in the previous domain have now been replaced by slower environmental processes. The spatio-temporal extent of this domain is significantly greater, and it can cover hundreds of kilometers and months of post-oil spill evolution.

Advanced Oil Spill Modeling Techniques

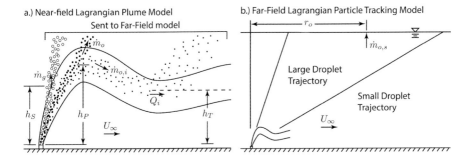

FIGURE 8.1 The two distinct domains of subsea blowout modeling. (a) On the left is the near-field domain, where the plume dynamics are modeled. Once the oil plume is considered neutrally buoyant, the simulation enters the far-field domain. (b) On the right, the far-field domain is where the oil is advected by the sea currents, the wind, and the waves (Socolofsky et al., 2015b).

8.3 SUBSEA BLOWOUTS AND NEAR-FIELD MODELING

8.3.1 INTRODUCTION

This chapter starts with the latest developments in 3D oil spill modeling, that is the modeling of a deep-water oil-well blowout and the subsequent creation of an oil plume. Subsea oil spills typically originate from the seafloor when the blowout preventer well head isolation mechanism fails to stop a sudden high-pressure gas shock wave. As was mentioned in the introduction, the research and development of this type of modeling were accelerated by two main factors. These are (1) the new direction that the oil industry has taken towards offshore deep-water exploration and (2) the Deepwater Horizon accident that highlighted important knowledge gaps on planning and responding to this type of oil spill. This chapter considers this type of modeling first because the outcomes of deep-water oil spill research have revealed important controls on the fate and circulation of oil spills in the marine environment that were not taken into account by previous far-field surface oil spill modeling. In other words, the newly acquired knowledge in oil-plume modeling has revealed some of the shortcomings of the far-field surface modeling as well.

One such example is the crucial role that oil droplet size plays in the fate and the circulation not only of the deep-water oil plume but also of the surface oil spills. Although the mechanisms that produce the oil droplets are different between a deep-water blowout and a surface oil spill, the droplet sizes are similar and some of the environmental processes acting on these droplets are the same. This fact also strengthens the argument that 3D modeling of surface oil spills in the marine environment is preferable over 2D-only surface oil spill modeling, as 2D modeling de-facto can't simulate the circulation of oil droplets in the water column as well as other 3D environmental processes that will be discussed later in this chapter.

8.3.2 Reservoir Fluid versus Surface Oil

The first major difference between subsea blowouts and surface oil spills is the nature of the fluid. In the first case the fluid comes through the oil well straight from the underground petroleum reservoir, which is the natural trap between different geological layers where the petroleum was stored. Due to the high reservoir pressure, a large amount of light hydrocarbons and other inorganic compounds are dissolved in the oil that would otherwise be found in the gas phase at atmospheric conditions. This pressurized oil is often called "live" oil, so as to distinguish it from "dead oil," which is the same oil after it has been depressurized to atmospheric pressure and temperature, during which process it loses much of the low molecular weight compounds. These dissolved gas compounds give the live oil certain characteristics such as lower density, lower viscosity, higher compressibility, and lower surface tension when compared to the dead oil at the same conditions. But the most important characteristic of the live oil is its phase behavior. It can undergo phase changes such as a transition from the liquid to the gas phase due to the pressure drop, or a transition from liquid to solid wax due to the temperature drop. Phase changes affect the buoyancy and mass transfer to the water column of the oil spill, thereby determining the fate and the trajectory of the oil particles. These changes are usually modeled with some variation of the equations of state, such as those of Vitu et al. (2006).

8.3.3 Modeling the Underwater Oil Jet

A number of factors associated with the oil properties, the oil-well conditions, the exit-pipe geometric characteristics, and the ambient subsea conditions determine the sizes of the formed oil droplets and gas bubbles. The sizes of these droplets and bubbles in turn determine the buoyancy and consequently the time that these particles will take to traverse the water column, and whether they will reach the sea surface (Figure 8.2). Furthermore, other processes, such as dissolution and biodegradation, reduce the mass of these particles, with rates that depend on their size. This is because these oil degradation processes act on the surface of the particles, and since smaller droplets have higher area-to-volume ratios, they degrade faster.

8.3.3.1 The Physical Processes

The first phase of a subsea blowout that requires simulation is the underwater oil jet that is formed when the well-head blowout preventer fails. This is a crucial stage of an oil spill, as it affects the fluid's breakage into droplets and bubbles as explained below. In turn, the size of these particles determines whether they will arrive at the sea surface and consequently the fate of the oil spill. The oil jet formed in a subsea blowout is characterized by high exit velocity and turbulent flow due to the pressure difference between seawater and the wellbore pipe coming from the underground petroleum reservoir (Figure 8.3). For a typical blowout scenario, the flow regime of the oil jet will be fully atomized due to turbulent breakup, according to the Ohnesorge (1936) classification system of flow regimes.

Advanced Oil Spill Modeling Techniques

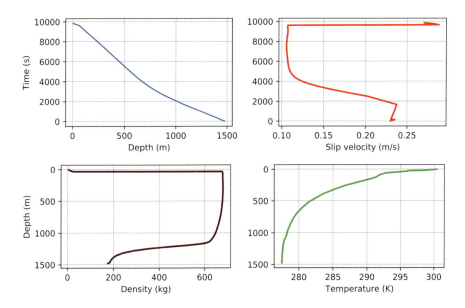

FIGURE 8.2 Oil spill bubbles can condense to droplets and back to bubbles as they ascend through the water column, drastically affecting their ascending (slip) velocity, density, and size. Condensation can happen due to the dissolution of light components into the seawater, while gas formation can happen due to pressure drop and rise in temperature.

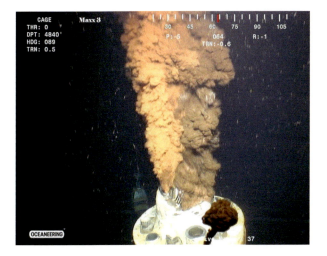

FIGURE 8.3 This turbulent oil jet was formed at 1500 m depth after the blowout preventer failed during the Deepwater Horizon accident in the Gulf of Mexico (McNutt et al., 2012).

The simulation of the oil jet is challenging due to the multiple physical processes and forces involved. In summary, the kinetic energy of the fluid increases as it accelerates inside the broken oil pipe, due to the pressure differential. The loss of pressure leads to expansion and possibly to formation of gas that accelerates the fluid further. At the exit point, the fluid experiences a rapid acceleration due to the sudden pressure drop, which can be in the order of hundreds of bars. The amount of gas present at this phase of the oil spill is crucial for controlling the turbulent kinetic energy of the jet flow and the subsequent size of the droplets and bubbles (Table 8.1, Figure 8.4). The breakage of the dispersed phase into droplets and bubbles is caused by two separate mechanisms. Initial breakage takes place at the outer edges of the jet due to the shear forces created by the different flow of the seawater and the oil jet. A secondary breakage of the initial droplets into smaller ones is caused by turbulence, when eddies of sufficient energy and appropriate size collide with these droplets. Finally, the sudden pressure drop of the fluid leads to phase changes such as the formation of bubbles inside the oil droplets that facilitate the breakage of the oil even further. The main antagonistic forces that determine the size distribution of the droplets and bubbles are the oil viscosity and oil–water interfacial tension that resist the breakage, while the shear forces created by turbulence and the gas formation inside the oil facilitate the breakage.

8.3.3.2 Current Modeling Approaches

To date there are two categories of models to predict the oil droplet and bubble size distribution created by the oil jet. Models of the first category, known as the equilibrium models, estimate the average particle size based on a balance between the antagonistic forces. These models employ dimensionless flow characterization numbers such as the Reynolds number (Re) and Weber number (We) to estimate the average particle size. The more recently developed population dynamic models estimate the droplet size evolution along the jet center line. These calculate both the probability of a droplet to break, based on the local turbulent energy dissipation rate, and the probability of a droplet to coalesce based on the population density of droplets.

TABLE 8.1

Combinations of Oil Types with Different Gas-to-Oil Ratios (GOR)

Oil Type	Depth (m)	GOR (ft³/bbl)	% Oil in the Water Column
Light	1500	1500	0.4
Light	500	1500	4.1
Light	100	1500	15.1
Medium	100	400	10.2

Note: The amount of gas present in the oil jet has a strong effect on the resulting droplet size distribution and the amount of oil that will reach the sea surface (Figure 8.4). Subsequently, the droplet size, density, and shape determine its buoyancy and the percentage of oil that stays in the water column.

Advanced Oil Spill Modeling Techniques

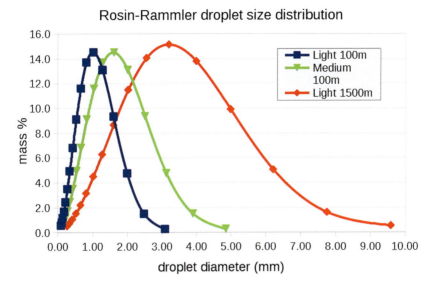

FIGURE 8.4 Effect of the presence of gas in the jet-flow on the droplet size distribution. Due to the lower pressure at 100 m depth, more gas escapes from the medium oil than from the light oil at 1500 m. As a result, the median droplet size of the medium oil is smaller than that of the light oil. In contrast, when at the same depth, the light oil produces smaller droplets.

8.3.3.3 Equilibrium Droplet Size Model

One of the most known models of this category is that of Johansen et al. (2013), which is an adaptation of the Wang and Calabrese (1986) model to oil jet experimental data which estimates the volume median droplet size d_{v50}. The d_{v50} is the size of the droplet where 50% of the oil volume has broken into smaller droplets and the other 50% of the oil has broken into larger droplets. d_{v50} is given by the mathematical relationship:

$$\left(\frac{d_{v50}}{DA}\right)^{\frac{5}{3}} = \frac{1}{We} + \frac{B}{Re} * \left(\frac{d_{v50}}{D}\right)^{\frac{1}{3}} \quad (8.1)$$

where A and B are empirical coefficients estimated from the experimental data, D is the oil jet exit diameter, Re is the Reynolds number equal to $(D\rho u)/\mu$, and We is the Weber number equal to $(D\rho u^2)/\sigma$, with u the initial jet flow velocity, ρ the oil density, μ the oil viscosity, and σ the oil–water interfacial tension.

Equation 8.1 is also used to calculate the median bubble size, with the gas properties replacing the oil properties (Figure 8.5).

The Weber number accounts for the interfacial tension forces and contributes more to the droplet size at lower jet flow velocities. In contrast, the Reynolds number accounts for the viscous and inertial forces and contributes more to the droplet size at higher jet flow velocities. In a typical subsea blowout preventer blowout, the ratio

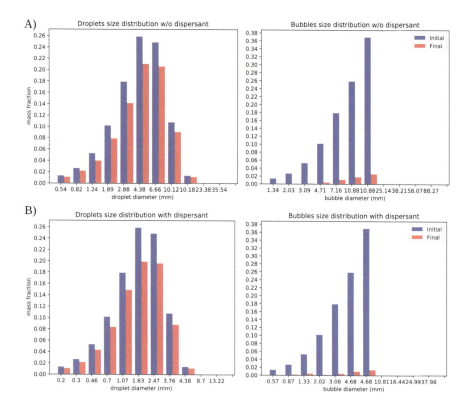

FIGURE 8.5 Droplet (left) and bubble (right) binned size distribution predicted by the Johansen et al. (2013) model, with (bottom) and without (top) chemical dispersant at 1500 m water depth for a light oil. The blue columns indicate the mass of each size at the jet, and the red columns represent the remaining mass at the surface due to dissolution into the water. Both droplets and bubble sizes are considerably smaller when dispersants are used. Bubbles are mostly dissolved in the water column.

of We/Re is small and interfacial forces control the droplet size. However, this ratio can have higher values when chemical dispersants are injected into the oil jet, since these can lower the interfacial tension by an order of magnitude. The Johansen et al. (2013) model also includes a jet velocity and jet buoyancy correction when there is a significant gas fraction in the flow (void fraction).

Once the median droplet size d_{v50} has been estimated, the complete droplet size distribution is derived with the use of a Rosin–Rammler distribution with an empirical spreading factor that is constant for all oils, although some modelers adjust the spreading factor according to the gas-to-oil ratio (GOR) of the fluid. The uncertainty of this approach lies in the empirical coefficients A and B of Equation 8.1, since they are estimated mainly from lab experiments (Brandvik et al., 2013) and are extrapolated to larger scale in real blowouts. Additionally, the constant spreading factor of the droplet size distribution does not always fit well the experimental data, especially for high-pressure "live" oils (Malone et al., 2018).

8.3.3.4 Population Dynamic Droplet Size Model

Zhao et al. (2014) combined a numerical population balance model together with a set of empirical correlations for jet hydrodynamics to predict the evolution of droplet size distribution along the oil-jet center line. The population droplet size model (VDROP-J) predicts the two competing rates. The breakage rate is based on the probability of a certain droplet size to collide with a sufficiently energetic eddy. In order to estimate the eddy energy, empirical hydrodynamic correlations are used to estimate the turbulent energy dissipation rate at specific distances on the jet's center line. The droplet coalescence rate is based on the probability of two droplets coming together. This probability is a function of the droplet population density, droplet size, and kinetic energy at the specific distance on the jet center line. Starting from an initial droplet size at the oil exit, the model estimates the droplet and bubble size distribution in small steps along the jet's center line (Figure 8.6). The advantage of this approach is that there is no need to extrapolate empirical lab correlations to a larger scale, as with the previous approach. Additionally, the size distribution of the droplets is directly calculated, in contrast to the previous approach where the spread of the statistical size distribution is fixed. Still, the uncertainty of this approach is the estimation of the turbulence dissipation rate from hydrodynamic correlations of single-phase jets. The effect of the multiphase flow and gas formation in the turbulent kinetic energy and dissipation rate is not yet fully understood, and is a topic of active research.

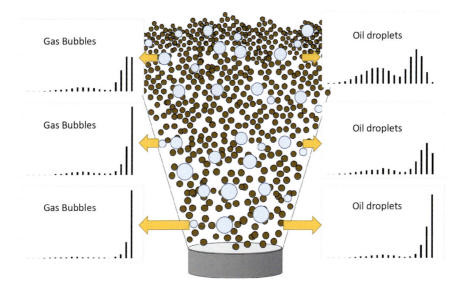

FIGURE 8.6 Graphical representation of the droplet and bubble size evolution along the jet trajectory (Zhao et al., 2014). Bubbles (dark gray) and droplets (light gray) start with all bubbles and all droplets having the same size, with the population gradually breaking down to both smaller bubbles and droplets due to turbulence and shear flow forces.

8.3.3.5 Modeling Oil Plume Dynamics and Underwater Degradation Processes

In a short distance from the oil exit, the velocity of the oil-jet drops and the flow of the dispersed fluid changes from jet-like to plume-like. The resulting plume contains liquid droplets, gas bubbles, entrained water, and potentially gas hydrates. Due to the initial jet flow and the physical properties of the dispersed phases, the plume exhibits complex dynamic behavior and moves inside the water column with its own momentum and buoyancy, in addition to being advected by the water currents. The initial stage of the plume is characterized by a sudden temperature and pressure drop as the fluid, coming from a hot, high-pressure underground reservoir, enters the water column. Figure 8.7 shows a typical phase envelope of a hydrocarbon mixture.

The sudden pressure drop can cause a liquid droplet to enter the dual-phase region of the envelope where both gas and liquid co-exist in the droplet. In other words, gas microbubbles will start appearing inside the droplet, and these will gradually increase in size as the droplet keeps ascending inside the water column. Likewise, when a gas bubble experiences a sudden temperature drop it may enter the dual-phase region of the envelope, with liquid forming inside the bubble. Additionally, gas hydrates can develop on the surface of the gas bubbles, forming a crust around the bubbles. This crust is made of tiny crystals in which water molecules trap small hydrocarbon molecules inside their crystal lattice. As can be seen on the phase diagram of methane hydrates (Figure 8.8), specific conditions are required for the creation of gas hydrates.

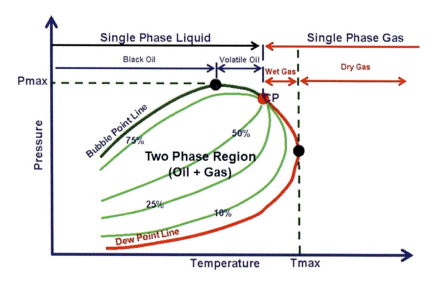

FIGURE 8.7 Typical phase envelope of a hydrocarbon mixture. Inside the envelope there is a two-phase region, in which percentages indicate the percentage of oil in an oil–gas mixture. Above the bubble point line there is only oil, and to the right of the dew point line only gas. High pressure and low temperatures favor oil, while the opposite conditions favor gas.

Advanced Oil Spill Modeling Techniques

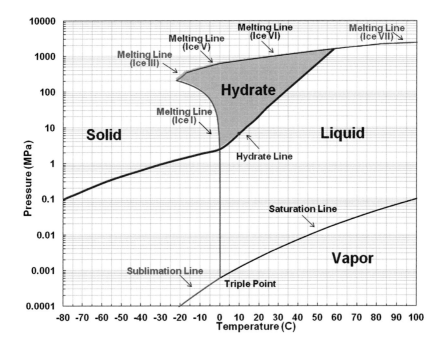

FIGURE 8.8 Methane hydrate phase diagram (source: ASTM Manual 73, Chapter 15). Low temperatures and high pressure (high depth) are required for methane hydrates to form. At 10°C, a pressure higher than 6.5 MPa is required to cross the hydrate line (red dot) into the hydrate formation envelope (gray area). This pressure corresponds to approximately 650 m depth in the seawater column.

As an example, at 10°C seawater temperature, a depth of more than 650 meters is required to enter the methane hydrate formation envelope (Figure 8.8). Additionally, only small hydrocarbon molecules can fit inside the cages of the crystal lattice, and as such the gas hydrates tend to form on the water–gas bubble interface. The gas hydrates change the behavior of the bubbles by reducing their buoyancy and greatly reducing the dissolution rate of gas compounds into the water.

This early stage is also characterized by high turbulent kinetic energy and momentum supplied by the oil jet. Both the oil jet and the initial stage of the plume are crucial for the droplet and bubble size distribution and the later circulation and fate of the plume particles. The jet-like flow quickly changes to a plume-like flow, and that transition depends on the characteristic plume length l_M (Zhao et al., 2016, and references within), which is equal to:

$$l_M = 0.94 \left(\frac{\rho D}{g \Delta \rho} \right)^{\frac{1}{2}} U_o \qquad (8.2)$$

where ρ is the discharged fluid density, $\Delta \rho$ is the density difference between the fluid and the seawater, D is the exit diameter, g is the gravity, and U_o is the exit velocity, with all units in the SI system.

The transition from jet to plume flow takes place between $l_M < x < 5l_M$, with Zhao et al. (2016) proposing $2l_M$ as the distance where most of the transitions take place. Zhao et al. (2016) also estimated that the turbulent energy dissipation rate remains almost constant to a distance of $10D$ (ten exit diameters), and then drops with distance.

As an example, in the case of the Deepwater Horizon blowout, for the given fluid density, exit diameter, and exit fluid velocity, the transition point from jet flow to plume flow, $2l_M$, is estimated at one meter from the exit when using Equation 8.2. The distance of constant turbulent energy dissipation rate, $10D$, is estimated at three meters from the exit according to Zhao et al. (2016). This finding means turbulence remains equally high at three times the distance of the jet flow, and well within the plume, and the breakage of the fluid into smaller and smaller droplets and bubbles continues well inside the plume.

As the plume rises in the water column, more seawater is gradually entrained in the plume. As a result, the plume's momentum is dispersed into a greater volume, the density difference with the seawater reduces, and so does its buoyancy, and the plume slows down. The plume also carries with it denser entrained water towards less dense shallower water, due to temperature and salinity differences in the water column (Figure 8.9). The plume may reach a certain depth where it will become naturally buoyant and be trapped at the trap height h_T. The plume may overshoot to a shallower depth due to its momentum, to the peak height h_P, and it also may retreat to a deeper depth than the trap height when the entrained denser water starts descending, dragging the plume with it. These forces create oscillations of the plume around the trap height (Figure 8.9). Eventually the plume settles at the trap height and starts moving passively with the water currents, which marks the end of the near-field domain.

The oil droplets and bubbles have their own ascending velocities (slip velocities) that may be different from that of the plume, and will depend on their size, density, and shape. When these particles are trapped inside the plume their trajectory is the sum of the plume's trajectory plus their movement inside the plume. Therefore, their ascending velocity may be higher than the velocity they have when they are free inside the water column, especially at the initial stages of the plume. One of the most accurate models to date to calculate the shape and ascending velocities of oil droplets and bubbles was developed by Zheng and Yapa (2000). In general, the bubble shape changes from spherical to elliptical to rounded cap as the size increases. Large droplets and bubbles can escape from the plume before the plume settles when there is strong current cross flow $U\infty$ (Figure 8.9), because of the large buoyancy difference between the two, further reducing the buoyancy of the remaining plume.

Other processes that need to be taken into account when simulating the near-field domain are the heat transfer rate, mass transfer rate, and the biodegradation rate between the oil particles and the water. All these rates are higher in smaller particles, because these processes take place on the water–particle interface, and smaller particles have a higher surface-to-volume ratio. These processes also affect the trajectories of the particles since they can cause phase transition and buoyancy change of the particles (Figures 8.10 and 8.11).

Advanced Oil Spill Modeling Techniques

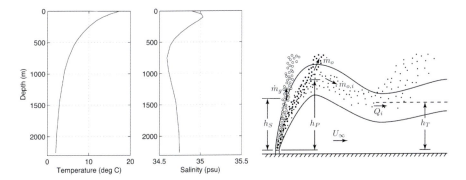

FIGURE 8.9 As the oil plume gradually rises, it enters less dense water, because the water temperature rises and the salinity generally drops with decreasing depth. Eventually, the plume becomes naturally buoyant at the trap height (intrusion depth) h_T, h_S = separation height of gas on the upstream side of the plume, h_P = the highest point along the plume trajectory, $\dot{m}_g$ = the mass fluxes of gas leaving the plume at the separation height, $\dot{m}_0$ = the mass fluxes of gas leaving the plume at the maximum height, $\dot{m}_{0,i}$ = the mass fluxes of gas entering the intrusion after the maximum height, $U\infty$ = ambient current velocity, and Qi = volume flux in the intrusion layer (modified from Socolofsky et al., 2015b).

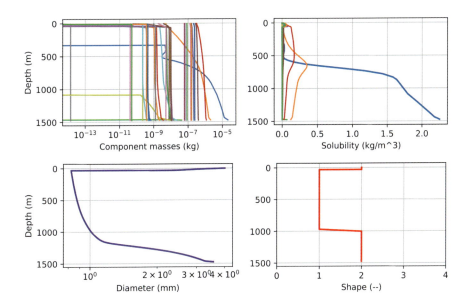

FIGURE 8.10 Preferential mass loss of the lighter bubble components due to their higher solubility. The heavier residue can condense and lose volume and buoyancy. The shape changes from type 2 (elliptical) to type 1 (spherical).

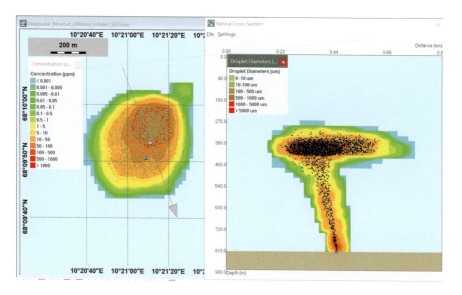

FIGURE 8.11 Example of plume simulation from OSCAR DeepBlow. The models track the plume as a multi-phase volume, including droplets, bubbles, dissolved compounds, entrained seawater, and gas hydrates. On the left, the color scale indicates different concentrations of dissolved compounds. On the right, one can see the plume reaching the trap height, forming an intrusion layer (Kotzakoulakis, 2020a)*. *Note: animation available in the references section.

The above short introduction of the processes taking place in the near-field domain of a subsea blowout highlights the complexity and the challenges that near-field models have to deal with. Some of the well-known commercial and open-source near-field models are listed in Table 8.2, together with their main characteristics.

These near-field models follow the same basic approach to calculating the evolution of their state-space variables along the center-line trajectory of the plume by solving conservation equations, in small time steps, for mass, momentum, energy, and concentration. These quantities are averaged over each plume cross-section for all the models except for the TAMOC model, which in stratified plume mode can simulate two entrainment layers for each cross-section. All the models except MOHIDJet also track the ascending velocities of the bubbles and droplets in relation to the plume, and their exit from the plume, as well as the mass transfer of oil compounds into the water column (Figure 8.11).

Finally, not all the models have the capability to use equations of state (EOS) to calculate phase changes as the bubbles and droplets ascend the water column. To date, TAMOC seems to have the most complete set of features as shown in Table 8.2, including EOS for the particles, two water entrainment layers, and furthermore it is offered as a free, open-source model to be used and extended with more features. In comparison, commercial models have better technical support and have easier to use graphical interfaces. An integrated, enhanced version of the TAMOC near-field model that is coupled with OpenDrift far-field has been developed by Kotzakoulakis

TABLE 8.2

Near-Field Oil Spill Models and Their Main Characteristics

Characteristics	BLOSOM	MIKE n.f.	MOHIDJet	OILMAP D.	OSCAR D.B.	TAMOC
Jet model	Yes	No	Yes	Yes	Yes	Yes
Plume model	Yes	Yes	Yes	Yes	Yes	Yes
Dispersed phases	**Multi-phase**	**Multi-phase**	Single-phase	**Multi-phase**	**Multi-phase**	**Multi-phase**
Entrainment layers	Single-layer	Single-layer	Single-layer	Single-layer	Single-layer	**Dual-layer**
Usage model	**Free to use**	Commercial	**Free to use**	Commercial	Commercial	**Free to use**
Code model	Closed source	Closed source	**Open source**	Closed source	Closed source	**Open source**
Reference	1	2	3	4	5	6

Sources: 1. Sim et al. (2015); 2. DHI MIKE; 3. Paiva et al. (2017); 4. RPS Oilmap Deep; 5. SINTEF OSCAR DeepBlow; 6. Socolofsky et al. (2015a).

240 Oil Spill Occurrence, Simulation, and Behavior

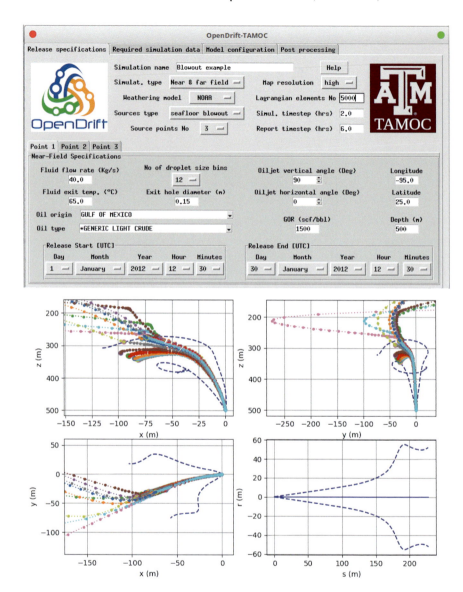

FIGURE 8.12 Part of the integrated OpenDrift-TAMOC model's user interface is shown at the top where the initial conditions of near-field and/or far-field simulation can be specified. An example of the near-field simulation is shown at the bottom where the outline of the plume is marked with dashed lines. Large droplets and bubbles escape from the plume earlier and follow trajectories that depend on their buoyancy and water column currents, while smaller ones remain inside the plume until the end of the near-field simulation. The modified integrated model (Kotzakoulakis et al., 2019) reads and updates the vertical water column profile frequently from the available hydrodynamic data.

Advanced Oil Spill Modeling Techniques

et al. (2019) that includes a graphical user interface and additional capabilities such as automated estimation of reservoir fluid composition based on the NOAA oil database and GOR data, automated vertical water column profile extraction and updating from hydrodynamic data, and automated transfer of near-field results to the far-field model (Figure 8.12). Additionally, the OpenDrift far-field model (Dagestad et al., 2018) has been enhanced with new capabilities, including surface oil spreading, an improved evaporation model, an oil dissolution model, a biodegradation model, and batch-run and statistical processing of multiple simulations for oil spill contingency planning.

8.3.3.6 Typical Near-Field Model Inputs and Produced Results

The typical input data for near-field models are (Figures 8.12 and 8.13):

- The vertical water column profile, which as a minimum includes at each depth the water current velocities, water temperature, and salinity.
- The reservoir fluid composition, either as a single-phase fluid or separate oil and gas compositions and GOR. Pseudocomponents are typically used for the oil composition in order to reduce the time of the required calculations and the complexity of the required data.
- Location data, including date-time, longitude, latitude, and depth of the oil release point(s).

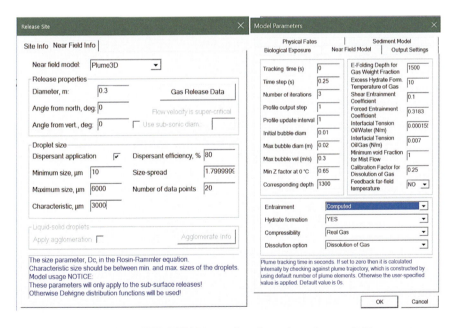

FIGURE 8.13 Part of SINTEF OSCAR's user interface where the near-field parameters can be specified. On the left are the oil jet and the droplet size parameters. On the right are the near-field simulation parameters.

- Jet initial conditions such as flow rate, flow duration, exit diameter, vertical inclination and horizontal direction, fluid temperature at the exit point, fluid phase(s) viscosity, and interfacial tension with the seawater.
- The number of bins (number of particle sizes) that the particle size distribution should be binned to for both liquid and gas.

A typical set of near-field results includes:

- Evolution of each particle's mass down to pseudocomponent level, including remaining component masses in the droplets and bubbles, and dissolved component masses and concentration in the water column (Figures 8.10 and 8.11).
- Evolution of each particles' properties including temperature, density, diameter, shape, and phase (Figures 8.2 and 8.10).
- Trajectory of each particle including position and slip (ascending) velocity (Figures 8.2 and 8.12).
- State-space results for the plume as a whole, including center-line position, orientation, width, momentum, and velocity (Figures 8.11 and 8.12).
- Mass balance of the dispersed fluid between the different environmental compartments, including oil and gas reaching the sea surface, compounds dissolved in the water column, oil particles in the water column, and oil sinking to the sea floor and being incorporated as a sediment (Figure 8.14).

At the end of the near-field simulation the complete set of results is stored and the final positions and masses of the particles and the dissolved components are passed to the far-field model.

8.4 FAR-FIELD OIL SPILL MODELING

The recent intense research activity after the Deepwater Horizon accident on the processes affecting the fate of oil spills has also contributed to the advancement of the far-field models. Some of the key environmental processes are now better understood and modeled with more advanced parametrizations. One such example is the dispersion of surface oil into the water column. Historically this process was estimated in 2D surface simulations with the use of empirical factors. Most of the far-field models are now simulating the process in 3D mode with the modeling of wave entrainment, the estimation of the formed droplet size distribution, and the circulation of the droplets in the water column.

Far-field simulations are performed for both surface oil spills and subsea oil leaks. A surface oil spill simulation typically starts by introducing a large number of surface elements (spillets) that together form the simulated oil spill. Each spillet has a set of initial properties such as mass, density, and thickness, and is usually composed of a certain number of pseudocomponents. Each pseudocomponent is a group of oil compounds with specific physical and chemical properties that together give the spillet its bulk properties. Each spillet is tracked independently, and its properties

Advanced Oil Spill Modeling Techniques

Predicted mass balance for DWH blowout, TAMOC near-field

A) Mass (%) without chemical dispersant

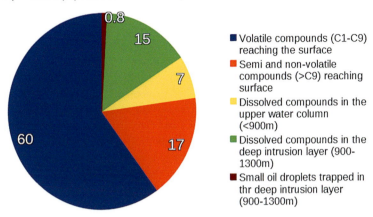

B) Mass (%) with application of chemical dispersant at the oil jet

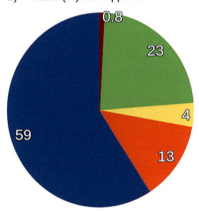

FIGURE 8.14 Prediction of the mass (%) distribution of the Deepwater Horizon blowout fluid between the different environmental compartments. At the top (A) is the prediction of the TAMOC near-field model for the Deepwater Horizon subsea blowout without the use of dispersant. At the bottom (B) is the same scenario, but with the injection of dispersants at the oil-jet exit (Gros et al., 2017).

can change over time due to environmental processes at the surface (weathering) or when entrained into the water column. 3D far-field models have an additional set of properties for each element such as droplet diameter, so as to facilitate simulation in the water column.

Some far-field models can also track the circulation and fate of dissolved oil compounds. This is usually done with Lagrangian elements that are advected passively by the water currents, carrying certain masses of the compound.

When the far-field model is coupled with a near-field model for the simulation of a subsea blowout, all the elements at the end of a near-field simulation are passed to the far-field model as independent particles since the plume has lost its own momentum and buoyancy (Figure 8.15). The near-field simulation will produce gas bubbles, oil droplets, sedimented oil, and dissolved compounds that make up the total of the released pressurized fluid (Figure 8.14). The most advanced models can continue tracking all these phases in the far-field domain, but as a minimum the most important phase that needs to be tracked is the oil droplets and oil spillets on the sea surface. The reason for this is that the extreme majority of the gas bubbles and a smaller quantity of the droplets will quickly either dissolve in the water column (Figures 8.5 and 8.10), or reach the surface and enter the atmosphere. The concentration of these dissolved compounds will quickly drop to non-detectable levels a short distance away from the origin in open sea (Figure 8.11). Concentration can be of concern when the origin is in shallow waters close to the shoreline, or in water bodies of restricted water renewal such as harbors, fjords, and lakes. Similarly, surfacing bubbles may be of concern at the surfacing location, but concentration of airborne compounds will quickly drop a short distance from that location.

8.4.1 Oil Spill Advection

An oil spill is advected by three forcing fields: the sea currents throughout the water column, and the wind and the waves on the surface. In addition, wave entrainment, eddy diffusivity, and droplet buoyancy have an additional effect on the circulation of droplets in the water column. Furthermore, changes in the properties of the oil due to weathering processes also affect the oil circulation. In addition to the oil properties, a far-field model requires as a minimum the surface water current velocity field, the wind velocity field at 10 meters above the surface, and the sea-surface temperatures as model inputs, with parametrizations available to estimate the remaining processes. For a 3D simulation, current velocities and temperatures at different depths are required. Additional inputs such as wave data are important in order to minimize the uncertainty of certain parametrizations, as will be discussed later. The far-field models can then calculate the trajectories of the introduced elements in small time steps, taking into account the forcing from all the available fields and solving the trajectories for the simulation period with a numerical technique such as the Euler method or the Runge–Kutta second- or fourth-order methods. In every time step the properties of each oil element, spillet, or droplet are updated with the effects of several environmental processes. Table 8.3 summarizes the features of some popular open-source and commercial models.

The predictions of the far-field models are sensitive to the model inputs, empirical factors, and parametrization techniques. As one would expect, the water currents have the greatest effect on the trajectories of the oil elements, followed by the wind and wave data as these are the main driving forces. Important factors to consider when selecting water current data are the type of data, the spatio-temporal resolution of the data, and the type of data grid. Although all hydrodynamic models derive their predictions from the Navier–Stokes equation, the way they divide the

Advanced Oil Spill Modeling Techniques

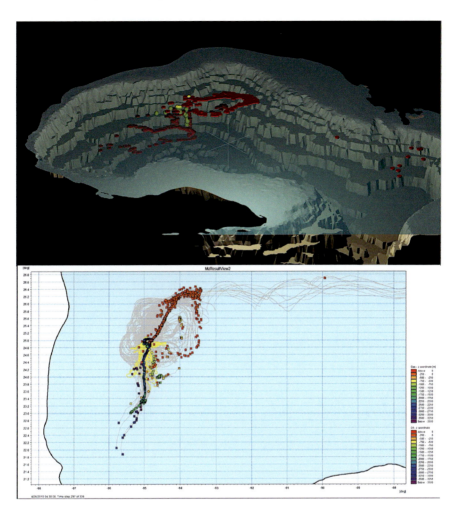

FIGURE 8.15 Example of a subsea blowout in the Gulf of Mexico with a coupled 3D far-field and near-field model (DHI MIKE model. Hørsholm. Denmark). In the top graphic, in yellow are the gas bubbles rising to the surface close to the release point. In red is the oil, part of it sinking to the sea bottom, another part staying in the water column, and a third part rising to the surface (Kotzakoulakis, 2020b)*. In the bottom graphic, one can see in yellow the gas bubble trajectories, in blue the underwater plume heading south, and in red the surface spill is heading to the north (Kotzakoulakis, 2020c)*. *Note: animation available in the references section.

three-dimensional water volume into layers and calculation cells affects the predictions of the models (Figure 8.16).

The three main categories of the hydrodynamic models are (1) the terrain following layer models (e.g. ROMS), (2) the horizontal layer models (e.g. NEMO), and (3) the hybrid models (e.g. HYCOM). One model may perform better than the others at different locations, depending on factors such as depth and sea floor morphology.

TABLE 8.3
Features of a Selection of Free, Open-Source, and Commercial Far-Field Oil Spill Models

Characteristics	OpenDrift-TAMOC[1]	PyGnome[2]	DHI MIKE	SINTEF OSCAR	RPS SIMAP
Graphical UI	Yes	Web interface	Yes	Yes	Yes
3D model	Yes	Yes	Yes	Yes	Yes
Forcing fields	Currents/wind/waves	Current/wind	Currents/wind/waves	Current/wind	Current/wind
Integrated with near-field	Yes (TAMOC)	Yes (TAMOC)	Yes (MIKE_jet)	Yes (DeepBlow)	Yes (Deep)
Weathering processes					
Spreading	Yes	Yes	Yes	Yes	Yes
Evaporation	Yes	Yes	Yes	Yes	Yes
Emulsification	Yes	Yes	Yes	Yes	Yes
Dispersion	Yes	Yes	Yes	Yes	Yes
Dissolution	Yes	Yes	Yes	Yes	Yes
Biodegradation	Yes	No	Yes	Yes	Yes
Sedimentation	No	Yes	Yes	Yes	Yes
Photo-oxidation	No	No	Yes	Yes	Yes
Response options	No	Yes (ROC)	Yes	Yes	Yes
Statistical modeling	Yes	No	Yes	Yes	Yes
Reference	1	2	3	4	5

Sources: 1. OpenDrift-TAMOC model at GitHub; 2. NOAA PyGnome model at Github; 3. DHI MIKE; 4. SINTEF OSCAR, 5. RPS SIMAP.

Notes:

[1]. OpenDrift-TAMOC is a custom model based on OpenDrift and TAMOC and developed by Kotzakoulakis et al. (2019). https://github.com/KKotzak/OpenDrift-TAMOC.

[2]. This is the beta version of PyGnome, with additional capabilities; the stable GNOME is 2D only.

Advanced Oil Spill Modeling Techniques

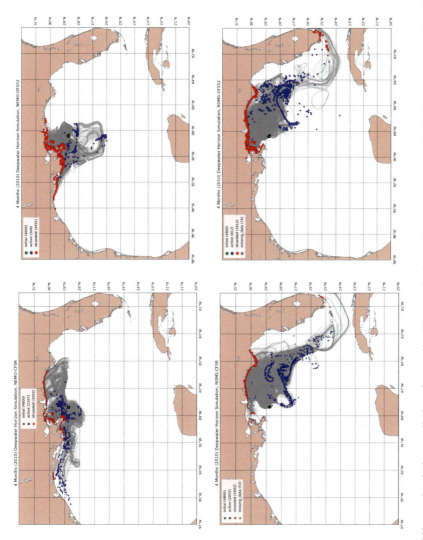

FIGURE 8.16 Different circulation predictions for an oil spill scenario due to the water current and wind inputs coming from different hydrodynamic and atmospheric models for the same period. Currents on the top panels are from the NEMO hydrodynamic model (www.nemo-ocean.eu), and from the ROMS on the bottom panels (www.myroms.org). Winds are from the CFSR atmospheric model (The climate data guide: climate forecast system reanalysis (CFSR)) on the left panels, and from the DFS5.2 on the right panels (Dussin et al., 2016).

When oil spill analysis or reanalysis data are available, this should be preferred to data produced from free model runs. Analysis and reanalysis data are produced by assimilating field measurements into the hydrodynamic models, in order to "tune" the model predictions with the field data and reduce the uncertainty compared to free model runs, where the hydrodynamic model predictions are based only on the initial and boundary conditions. Typical cases where reanalysis data can be used are statistical studies of oil spill scenarios in order to estimate the probability of certain areas to be hit by an oil spill. These studies perform multiple simulations in the same region and use many years of hydrodynamic data produced from reanalysis. In contrast, operational oil spill models are used in real-time incidents in order to assist the response authorities to minimize the impact of an oil spill. These models produce forecasts for the next few days and use forcing data (currents, winds, waves) exclusively from hydrodynamic and atmospheric model simulations (Figures 8.17 and 8.18), since field measurements do not yet exist. As the event unfolds over the following days, the oil spill models are updated with field data on the current location and state of the spill, in order to correct the previous predictions and provide forecasts for the following days.

Forcings

Show variables of no groups define ∨   Level   no levels defined

No.	Description	Type	Value		File Name
1	Horisontal drift speed	Built-in			...
2	Horizontal drift direction	Built-in			...
3	Vertical drift speed	Built-in			...
4	Wind speed	Constant	2	m/s	...
5	Water Temperature	Constant	10	degrees c	...
6	Water Salinity	Constant	0	psu	...
7	Solar radiation	Constant	1000	watt/m2	...
8	Wave height	Constant	0	m	...
9	Wave period (T02)	Constant	5	s	...
10	Area map with probability	Constant	1	probability	...
11	Surface elevation	Built-in			...
12	Volume of grid cell	Built-in			...
13	Area of grid cell	Built-in			...
14	Icefraction	Constant	0	dim. less	...
15	Ice thickness	Constant	0	m	...
16	Ice drift, u vector	Constant	0	m/s	...
17	Ice drift,v vector	Constant	0	m/s	...
18	Probability map of a parti	Constant	0	dim. less	...
19	Current u component	Constant	0	m/s	...
20	Current v component	Constant	0	m/s	...

FIGURE 8.17 Part of the DHI MIKE user interface for the specification of the forcing data for modeling an oil spill. Advanced forcing options such as detailed wave characteristics and solar radiation provide detailed data for the modeling of the oil spill processes that are affected by them, such as oil entrainment and oil photo-oxidation.

Advanced Oil Spill Modeling Techniques

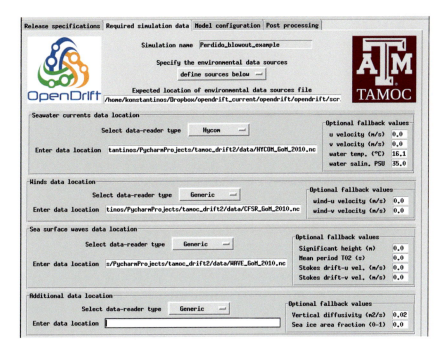

FIGURE 8.18 Part of the OpenDrift-TAMOC user interface for the specification of forcing data and other environmental conditions. Because OpenDrift is a general purpose Lagrangian model, it provides a variety of flexible data readers that are compatible with numerous types of forcing data.

8.4.2 THE IMPORTANCE OF WAVE DATA

It is clear from the required inputs of the aforementioned oil entrainment models that wave data that describe the height, the period, and the surface coverage with breaking waves are important for a reliable prediction of the entrainment, not only for the calculation of the oil entrainment rate and the oil intrusion depth into the water column, but also for the oil breakage that determines the size of the formed oil droplets and whether they stay in the water column. In many cases, wave data are overlooked and oil spill models rely solely on wind data to parametrize the aforementioned wave characteristics for the modeling of the oil entrainment, and droplet size distribution, as well as for the simulation of the drift generated by wave action, namely the Stokes drift. For the latter, in the absence of wave forcing data (Stokes drift velocity fields), or when the oil spill model does not have the capability to process wave data, an empirical factor is applied to the wind velocity that is used in order to account for the Stokes drift. This empirical factor is typically 2% for the wind advection and 1.5% for the Stokes drift, making a total of 3.5% of the wind velocity (Schwartzberg, 1971, Jones et al., 2016).

8.4.3 OTHER CRITICAL INPUTS AND PROCESSES

Excluding the forcing data that are directly driving the circulation of the oil spills, past sensitivity studies have identified the most critical inputs to the model predictions

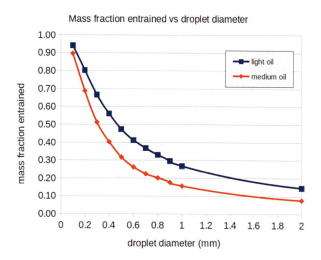

FIGURE 8.19 Example of the oil mass fraction entrained into a water column based on the droplet size and the oil type. The parametrized oil-entrainment due to breaking waves is based on a 7 m/s constant wind velocity. The medium oil is less entrained than the light oil, due to its higher viscosity and interfacial tension with the water.

(French-McCay et al., 2019; North et al., 2015). All studies agree that the droplet size distribution is probably the most crucial input for the fate of an oil spill. This is true not only for subsea blowouts, as discussed in the near-field modeling section, but also for surface oil spills when breaking waves entrain the oil into the water column, and oil droplets of various sizes can be formed based on the wave characteristics, the turbulent diffusion in the water column, and the properties of the oil. This is an area of active research due to the complexity of the processes involved and the importance of oil entrainment and droplet size distribution (Figures 8.19, 8.20).

8.4.3.1 Oil Entrainment and Droplet Size Distribution

As with the calculation of the subsea oil-jet droplet sizes, the models for the calculation of oil droplet sizes created by wave entrainment fall under the same two categories. In the first category are the models that use dimensionless flow characterization numbers (Johansen et al., 2015, Li et al., 2017), and in the second category are the population dynamics models (Nissanka and Yapa, 2017). The principles are the same as with the oil-jet models, and the interested reader is encouraged to review the referenced papers. For the calculation of the oil entrainment rate, Li et al. (2017) provide an updated method based on the dimensionless flow characterization numbers, the surface fraction covered by the breaking waves, and the wave height (significant height), together with the oil properties, the oil spill thickness, and empirical factors derived from experimental data. The Nissanka and Yapa (2017) calculation of the entrainment rate is based on the Tkalich and Chan (2002) formulation, and requires knowledge of the energy dissipation rate, with the authors providing an updated method for this calculation. The Nissanka and Yapa (2017) method also requires the

Advanced Oil Spill Modeling Techniques

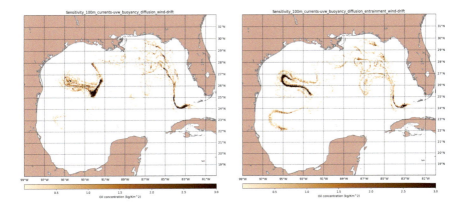

FIGURE 8.20 Example of the effect of droplet size on a far-field oil spill prediction. Both images are from the same oil spill simulation, but for two different droplet size groups. On the left are the larger size droplets (2 mm diameter), and on the right are the smaller size droplets (0.3 mm diameter) (Kotzakoulakis, 2020d)*. *Note: animation available in the references section.

wave characteristics (breaking wave coverage, wave height, wave period, wave type) as well as the oil properties and oil spill thickness.

8.4.3.2 Oil Spill Weathering and Oil Decaying Processes

The term weathering refers to changes of the oil's physical and chemical properties when it is exposed to the environmental conditions on the (sea) surface. The weathering processes can either decrease or increase the mass and volume of the oil spill, and they can affect oil spill trajectory and the effectiveness of response measures. The main weathering processes are:

1. *Spreading.* Although spreading of oil on the sea surface does not change any physical or chemical properties of the oil, any change in oil spill thickness and the area-to-volume ratio due to oil spill spreading affects the rate of all the other weathering processes, as well as the ability to recover and treat the oil spill. Therefore, accurate oil spill spreading modeling is important. Spreading is described by three regimes: gravity-inertia, gravity-viscous, and interfacial tension-viscous. The gravity-viscous regime is the most relevant in oil spill modeling, with viscosity being the most important oil property. At a later stage, other processes will produce uneven spreading. Wave entrainment will cause elongated shapes, with thicker oil spill downwind. Wind- and wave-induced Langmuir cells can create oil spill stripes parallel to the wind direction, and eddy diffusion can further spread and break the oil spill into patches. Most of the spreading models focus on the symmetrical spreading described by the gravity-viscous regime, but recent advances in oil spill spreading describe the latter processes (Galt and Overstreet, 2011).

2. *Evaporation* is the process responsible for the majority of mass loss during the first few days of a surface oil spill (Figure 8.21). The rate of evaporation is proportional to the vapor pressure of the oil and the oil spill surface area, among other factors. As the oil components have different fugacities, evaporation is a selective process that first removes the most volatile components. As the lighter components are gradually depleted, the evaporation rate reduces and practically stops after a few days. The residual oil will have higher viscosity, density, and lower buoyancy. The evaporation mass transfer coefficient for light oils depends mainly on the wind velocity and

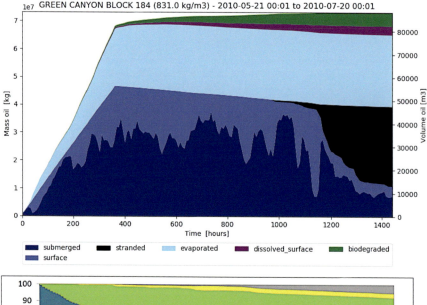

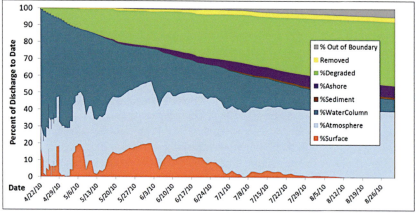

FIGURE 8.21 Examples of oil spill mass budgets from OpenDrift at the top and SIMAP at the bottom (SIMAP graph source: French-McCay et al., 2018 with permission). Evaporation is a fast-acting process, while other processes such as biodegradation can also have a significant impact, but at a later time.

Advanced Oil Spill Modeling Techniques

the oil/water temperature. In medium and heavy oils, composition gradients are formed between the surface and the inner mass of the oil due to the high viscosity and slow diffusion/mixing of the oil mass. The transfer coefficient of these oils is barely affected by the wind velocity, but instead depends on the mass renewal rate at the oil surface (Kotzakoulakis and George, 2018). The most common model approach is the grouping of the oil components in a few large "pseudocomponent" groups with specific vapor pressure and other physical properties (Figure 8.22).

3. *Emulsification* is the process where seawater is being trapped inside the oil spill body as suspended droplets. The mixing energy is provided by the waves and the wind, and the stability of the suspension depends on the composition and physical properties of the oil. If the weather conditions provide enough mixing energy such as breaking waves, a water-in-oil emulsion is formed rapidly during the first hours and days of an oil spill release (Figure 8.23). Emulsification can increase the volume of an oil spill by a factor of three, while it can also increase its viscosity by multiple orders of magnitude. These changes can make the chemical treatment and mechanical recovery of an oil spill challenging, and for this reason the response authorities need to be able to predict the behavior of the specific oil. One of the most popular and successful emulsification models is that of Fingas and Fieldhouse (2011), which is based on the calculation of the stability index. The stability index depends on a combination of density, viscosity, polar compounds, and asphaltenes. The model can provide an estimation of whether the oil will form an emulsion, the type of emulsion (unstable,

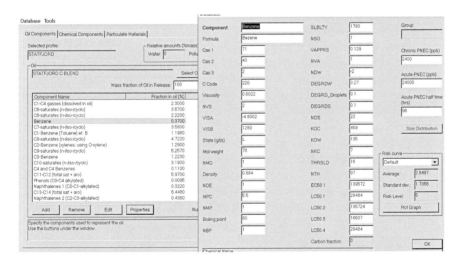

FIGURE 8.22 Part of the SINTEF OSCAR user interface for the specification of the oil composition and oil component properties. On the left, the oil is broken into several pseudo-components, or groups of components with similar properties. On the right, a detailed set of each pseudocomponent's properties can be defined.

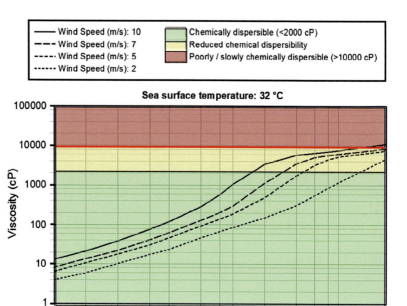

FIGURE 8.23 Effect of emulsification on the viscosity of the MC252 crude oil and the potential efficiency of chemical dispersants (Daling et al., 2014). As emulsification proceeds, chemical dispersants become less effective (yellow zone) and eventually have very poor effectiveness (red zone). The viscosity and dispersibility prediction presented here is part of the SINTEF OSCAR model.

meso-stable, stable), the percentage of water in the oil, and the required time to form the emulsion given specific weather conditions.
4. *Dissolution* is a slower process in comparison with the aforementioned ones. The rate of dissolution is mainly affected by the water solubility of the oil compounds, the contact area with the water, and the dissolution transfer coefficient across the boundary layer at the water–oil interface. Smaller oil molecules with some polarity such as light aromatic hydrocarbons have higher water solubility than n-alkanes with the same amount of carbon, so dissolution is more relevant for lighter oils with higher aromatic hydrocarbon content. In surface oil spills, dissolution has a small contribution to the overall weathering of the oil, because the very light and aromatic components are a small proportion of the whole oil, and they are also volatile and so tend to evaporate faster than they dissolve. In subsea blowouts, dissolution is important because the pressurized oil and any gas that is present contain high amounts of very light components. As an example, it has been calculated that during the Deepwater Horizon accident, 27% (and 22%) of the mass of the pressurized oil released underwater was dissolved into the water column with (and without) application of chemical dispersants

Advanced Oil Spill Modeling Techniques

(Gros et al., 2017) (Figure 8.14). Another case where dissolution may be important is in shallow waters with a slow water renewal rate, where toxic aromatic hydrocarbon concentrations can reach higher levels.

5. *Biodegradation* is also a slow process, but it can account for a large oil spill mass loss over a prolonged period of time, and is therefore an important process. It has been found that oil degrading bacteria exist in the ocean at multiple depths, even in locations that were never polluted by oil spills, but in very small numbers (Xu et al., 2018). When they are exposed to oil, various types of bacteria that consume specific groups of oil compounds will multiply exponentially until they become limited by the oil concentration, the availability of nutrients in the water, or the availability of oxygen dissolved in the water. The metabolic products of biodegradation are usually more water soluble, and these dissolve in the water column. Biodegradation takes place only at the oil–water interface, or in the water when hydrocarbons are dissolved. Therefore, entrained oil is more bioavailable than oil in a surface slick, and smaller droplets are biodegraded faster than larger droplets due to their higher area-to-volume ratio. Chemical dispersants help break oil into smaller droplets and can accelerate biodegradation. Oil droplets that lose volume to biodegradation also lose buoyancy, and they may never return to the surface (Figure 8.24). Biodegradation is typically modeled by a first-order decay equation with specific constants for each oil-group, bacteria-type combination.

6. *Sedimentation.* The majority of crude oils have a lower density than seawater, and they float when unweathered. Certain processes that take place in the marine environment can make them sink to the sea floor. Weathering processes such as evaporation, dissolution, and biodegradation preferentially remove the lighter components of the oil, leaving behind a denser oil spill residue. Suspended denser particles in the water column can attach

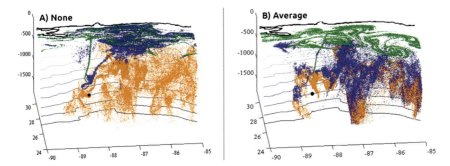

FIGURE 8.24 Effect of biodegradation on the buoyancy of oil droplets (North et al., 2015) in a vertical profile through a water column. In green are the large droplets, in blue are the medium droplets, and in brown are the small droplets. On the left (A) for an oil spill without biodegradation, both large and medium droplets rise to the surface, and the small droplets spread out at all depths. On the right (B) for an oil spill with an average biodegradation rate, only the large droplets remain at the surface, while most of the medium and small droplets remain suspended in the water column.

256   Oil Spill Occurrence, Simulation, and Behavior

onto the entrained oil droplets and make them sink. This is especially the case in shallow waters where inorganic particles from the seabed or from a river estuary are high in concentration. Some response options such as in-situ burning of the oil spill leave a denser residue that may sink. In deeper waters, the particulate matter concentration is lower, and the percentage of the oil spill that may sink is small.

7. *Photo-oxidation* is an oil weathering process with a slow rate. Because the oxidation energy is typically provided by sunlight, the oil needs to be near or on the sea surface. Viscous and opaque oils oxidize on the surface that is in contact with atmospheric oxygen, and together with other weathering processes create a hard surface layer that reduces further weathering. Lighter oils can create light-penetrable sheens, but other weathering processes act with faster rates and the contribution from photo-oxidation is small. Modeling of photo-oxidation is typically done with a first-order decay equation that depends on the light/UV intensity.

8.4.4 MODELING OF THE RESPONSE MEASURES

Most of the oil spill models include the means to make estimations about the effect of the different response options on a specific oil spill scenario. Each response option affects the oil spill in a different way, and, depending on the oil properties, the location of the oil spill, and the environmental conditions, a certain option may even have a negative environmental impact. The three main response options are the following:

- *Chemical dispersants*, which reduce the interfacial tension and to a certain degree the viscosity of the oil, enabling the oil spill to break into smaller droplets that may entrain more easily into the water column. Models take into account the altered oil properties to estimate the effect of each process they simulate, as discussed earlier.
- *Mechanical recovery* can also be modeled by specifying (1) the locations, (2) the rates of removal, and (3) possible oil spill containment barriers. The models take into account these three parameters in order to predict their effect on oil circulation, and the fate of the oil spill.
- In-situ burning of oil spills is modeled in a similar way to mechanical recovery, by specifying the burning rate and location. The degree of oil spill spreading and residual thickness modeling are important in the estimation of oil spill burning efficiency, because the burning will cease when the oil spill thickness reaches a minimum amount. Other processes are also affected by in-situ burning, due to the different properties of the remaining residue, such as the sedimentation process discussed earlier.

8.4.5 STATISTICAL ANALYSIS FOR CONTINGENCY PLANNING

In regions with petroleum exploration, production, and transportation activity, environmental agencies and response authorities need to be prepared and have a

Advanced Oil Spill Modeling Techniques

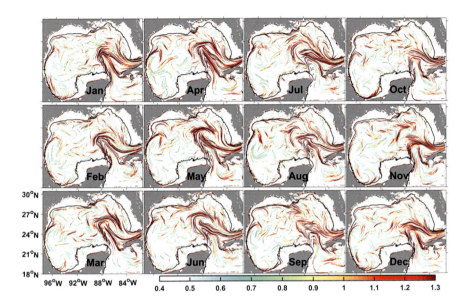

FIGURE 8.25 Climatological surface flow patterns by month for the Gulf of Mexico. The colored lines mark confluence maxima of flow trajectories. The black line marks the 50 m isobath. The climatological flow patterns were constructed by integrating 7-day flow trajectories for a period of 12 years (2003–2014) (Duran et al., 2018).

contingency plan to respond to an oil spill event. In order to do that, they need to identify the most probable origins of an oil spill, the potential duration, the amount of spilled oil, and the probable affected areas based on the production locations and transportation routes. This is possibly the most important application of oil spill models. The output of these simulations combined with sensitivity maps based on local ecosystems and human activities can be used to assess the risk for each location in the region of study. Oceans are characterized by large-scale and meso-scale circulation patterns in the order of hundreds of kilometers that can persist for weeks (Figure 8.25).

At the same time there are both seasonal variabilities and inter-annual variabilities, as well as sub meso-scale chaotic circulation, that make it difficult to predict the trajectory of a single oil spill event (Figure 8.26). Additionally, oil spills are advected by wind and waves, and their trajectories are affected by vertical mixing. For these reasons, oil spill simulations (scenarios) are performed that cover a large number of years, starting from the most probable oil spill location and with a shifting start date by a few days until the season(s) of interest is covered for all the simulated years (Figure 8.27).

The whole area of study, including the shoreline, is then split into a grid of geographic cells (bins), and the complete set of simulation results is statistically processed (Figure 8.28) in order to calculate statistical values for every geographic cell (Figure 8.27), such as the probability of oil spill arrival (Figure 8.29), the median oil concentration, the first arrival time, the day of maximum oil spill mass arrival, and

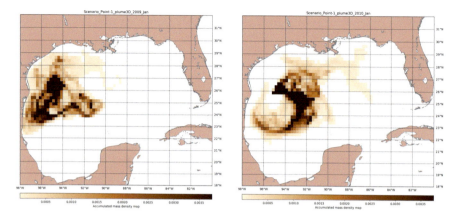

FIGURE 8.26 Example of the effect of inter-annual variability on the circulation of an oil spill. Both images are from the same scenario in the Gulf of Mexico, and use water current and wind data originating from the same models. The left simulation is for 2009, while the right is for 2010.

so on. The statistical results can then be combined with specific sensitivity information for each cell location, based on the local ecosystems (Figure 8.30) and human activity in those cells, in order to assess the risk and impact from a possible oil spill. These risk maps can be calculated for different seasons based on the specific seasonal circulation patterns, or for specific months.

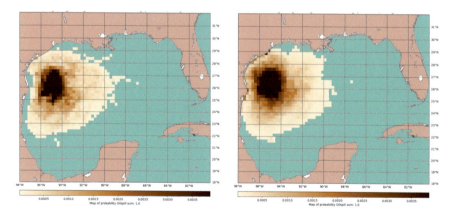

FIGURE 8.27 Example of probability maps produced by OpenDrift-TAMOC from 20 simulations of 2-month periods, starting on the January 1 for 20 consecutive years (1993–2012). The same scenario was performed in 3D mode (left), and in 2D mode (right). The 2D mode cannot predict the effect of wave entrainment on oil spill circulation, and the wind drove the oil spill to the shore. In the 3D simulation the oil spill does not reach the shore, since part of the oil spill was entrained and so was not advected by the wind.

Advanced Oil Spill Modeling Techniques 259

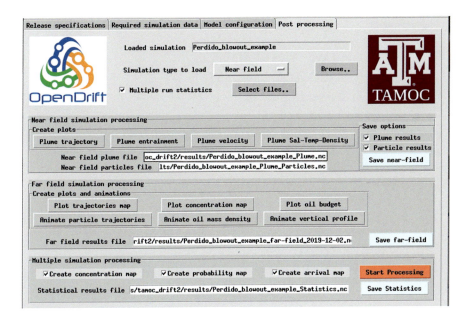

FIGURE 8.28 Part of the OpenDrift-TAMOC post-processing user interface. A single or multiple saved simulation can be loaded in order to produce trajectory maps, oil concentration maps, or statistical maps from multiple simulations.

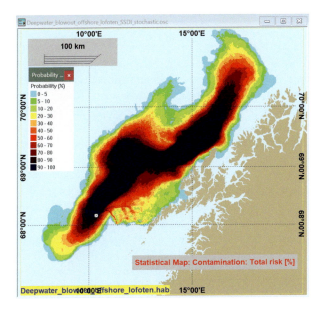

FIGURE 8.29 Example of a probability map produced by SINTEF OSCAR for an off-shore area close to Norway. The total risk is the probability of the oil spill reaching a specific location, which takes into account both the underwater plume and the surface oil spill circulation.

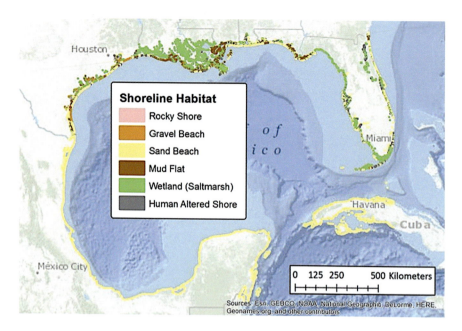

FIGURE 8.30 Example of the local ecosystems on the shorelines of the Gulf of Mexico. Different sensitivity ratings can be assigned to each ecosystem, and when combined with the probability and concentration maps produced by the oil spill simulations can be used to estimate the impact on each ecosystem of a specific oil spill scenario (source: Galagan et al., 2018).

8.5 SUMMARY

It is evident from the advances in oil spill simulators discussed in this chapter that the field of oil spill modeling has experienced increased research activity during the past decade. This was partly due to the response of the research community and the civil authorities to the gap in knowledge and the associated risks revealed by the Deepwater Horizon accident, and the general trend for oil exploration in deeper waters. New advanced parametrizations have been developed for key processes in both the near-field plume models and the far-field domain, such as population models for the breakage of oil into droplets, the plume dynamics, the phase changes of the pressurized oil, and the wave-induced entrainment of the oil. The examples presented herein are based on a selection of some of the most modern and popular simulators available for both the academic community and the commercial industry. Although the feature sets of the models shown in Tables 8.2 and 8.3 are similar, there are a few distinct differences based on their origin and purpose. Open-source models such as TAMOC, OpenDrift, and GNOME are developed by the academic community and public institutes, and are offered free to use and to customize. This ability makes them ideal for researchers wanting to experiment with new parametrizations, and students wanting to learn the inner workings of a simulator. In contrast, commercial models such as MIKE and OSCAR involve a cost to use but provide state-of-the-art

Advanced Oil Spill Modeling Techniques

parametrizations, a detailed graphical interface, technical support, and modeling services. These characteristics make them the better choice for industry clients and decision-making authorities. However, it is important to note the big advances that OpenDrift, GNOME, and TAMOC have achieved in the last few years, which have closed the gap to the commercial models in features and in state-of-the-art process parametrizations.

Another distinction between the presented models is their original intent and therefore their strongest points. Both OpenDrift and MIKE are general-purpose models with specific modules for oil spill modeling. As such, they have more flexibility and variety in the type of data that they can utilize. For example, as shown in Figures 8.17 and 8.18, OpenDrift and MIKE include detailed wave data that give a better description of the forcing fields, and as such they can better predict the circulation of an oil spill. Additionally, MIKE is integrated with a hydrodynamic model and a wave model that can produce the required data for a specific location and in fine detail in cases such as complex coastal regions, by utilizing unstructured grids. In contrast, purpose-built oil spill models such as OSCAR have fewer options for forcing inputs, but retain an advantage with better description of the oil chemistry as shown in Figure 8.22, and have more sophisticated weathering algorithms that better describe the changes that the oil undergoes in the marine environment.

There is a two-way relationship between the state of the oil spill and its circulation. For example, the emulsification of oil affects the amount of oil entrained in the water column, and the amount of oil evaporated. Therefore, advanced oil spill modeling requires very accurate description of both forcing fields (currents, winds, waves, turbulence) and environmental conditions and processes that alter the properties of the oil.

A typical set of near-field model inputs and produced results is given in Section 8.3.3.6. Typical inputs of a far-field include oil leak location, oil type, leak rate, and leak duration of the oil spill as well as the environmental conditions (temperature, salinity, ice coverage) and the forcing fields (winds, water currents, waves). Typical results from the far-field model include the mass budget of the oil spill that indicates the mass associated with each environmental process and compartment (sea surface, water column, sediment, and atmosphere), oil trajectory maps, and oil concentration maps. Post-processing of multiple far-field simulations can produce oil spill hit probability maps, oil spill arrival time maps, and risk maps. The results are stored in a standardized file format such as the netCDF file format so they can be read and processed by other software tools for further analysis.

8.6 ACKNOWLEDGMENTS

Support for this work was provided by the Consejo Nacional de Ciencia y Tecnología (CONACyT) – Secretaría de Energía (SENER) Grant 201441, as part of the Consorcio de Investigación del Golfo de México (CIGoM).

We also wish to thank:

Dr. Julio Pardo Sheinbaum and Dr. Paula Pérez Brunius from CICESE Physical Oceanography for their guidance and useful advice.

MIKE Powered by DHI for providing the MIKE Zero with a full license for the needs of the chapter, and particularly Mikkel Andersen, Christian Appendini, and Rafael Meza-Padilla for their help and support with setting up and running the MIKE Zero model.

SINTEF Ocean for providing the OSCAR model with a full license for the needs of the chapter and particularly Jørgen Skancke for providing the OSCAR scenarios and May Kristin Ditlevsen for providing support for the model.

The scientific open-source community and particularly the creators of the TAMOC and OpenDrift models for providing these models to the community free to use and modify, which freedoms have been used in this chapter.

REFERENCES

Brandvik, P.J., Ø. Johansen, F. Leirvik, U. Farooq and P.S. Daling. 2013. Droplet breakup in subsurface oil releases–Part 1: Experimental study of droplet breakup and effectiveness of dispersant injection. *Marine Pollution Bulletin*, 73(1), pp. 319–326.

Camilli, R., C.M. Reddy, D.R. Yoerger, B.A.S. Van Mooy, M.V. Jakuba, J.C. Kinsey, C.P. McIntyre, S.P. Sylva and J.V. Maloney. 2010. Tracking hydrocarbon plume transport and biodegradation at Deepwater Horizon. *Science*, 330, pp. 201–204.

Dagestad, K.F., J. Röhrs, Ø. Breivik andB. Ådlandsvik. 2018. OpenDrift v1. 0: A generic framework for trajectory modelling. *Geoscience Model Development*, 11, pp. 1405–1420. doi:10.5194.

Daling, P.S., F. Leirvik, I.K. Almås, P.J. Brandvik, B.H. Hansen, A. Lewis and M. Reed. 2014. Surface weathering and dispersibility of MC252 crude oil. *Marine Pollution Bulletin*, 87(1–2), pp. 300–310.

DHI MIKE model. *Hørsholm*. Denmark. https://www.mikepoweredbydhi.com/

Duran, R., F.J. Beron-Vera and M.J. Olascoaga. 2018. Extracting quasi-steady Lagrangian transport patterns from the ocean circulation: An application to the Gulf of Mexico. *Scientific Reports*, 8, p. 5218.

Dussin, R., B. Barnier, L. Brodeau and J.M. Molines. 2016. The making of the Drakkar forcing set DFS5. *DRAKKAR/MyOcean Report*, pp. 1–4.

Fingas, M.F. and B. Fieldhouse. 2011. Water-in-oil emulsions: Formation and prediction. *Proceedings of the 34th Arctic and marine oil spill program technical seminar.* https://inis.iaea.org/search/search.aspx?orig_q=RN:43033188

French McCay, D., M. Horn, Z. Li, D. Crowley, M. Spaulding, D. Mendelsohn, K. Jayko, Y. Kim, T. Isaji, J. Fontenault, R. Shmookler, and J. Rowe. 2018. Simulation Modeling of Ocean Circulation and Oil Spills in the Gulf of Mexico. Volume III: Data Collection, Analysis and Model Validation. US Department of the Interior, Bureau of Ocean Energy Management, Gulf of Mexico OCS Region, New Orleans, LA. OCS Study BOEM 2018-041; 313 p. Obligation No.: M11PC00028. https://espis.boem.gov/final reports/BOEM_2018-041.pdf.

French-McCay, D., D. Crowley and L. McStay. 2019. Sensitivity of modeled oil fate and exposure from a subsea blowout to oil droplet sizes, depth, dispersant use, and degradation rates. *Marine Pollution Bulletin*, 146, pp. 779–793.

Galagan, C.W., D. French-McCay, J. Rowe, L. McStay, and D. Crowley. 2018. Simulation modeling of ocean circulation and oil spills in the Gulf of Mexico. Volume I: Synthesis report. U.S. Department of the Interior, Bureau of Ocean Energy Management, Gulf of Mexico OCS Region, New Orleans, LA. OCS Study BOEM 2018-039. 164 p. Obligation No.: M11PC00028. https://espis.boem.gov/technical summaries/BOEM_2018-039.pdf

Advanced Oil Spill Modeling Techniques

Galt, J.A. and R. Overstreet. 2011. *Development of spreading algorithms for the response options calculator (ROC), by Genwest Systems, Inc.* (Genwest Technical Publication 11-003).

Gros, J., S.A. Socolofsky, A.L. Dissanayake, I. Jun, L. Zhao, M.C. Boufadel, C.M. Reddy and J.S. Arey. 2017. Petroleum dynamics in the sea and influence of subsea dispersant injection during Deepwater Horizon. *Proceedings of the National Academy of Sciences*, 114(38), pp. 10065–10070.

Hazen, T.C., E.A. Dubinsky, T.Z. DeSantis, G.L. Andersen, Y.M. Piceno, N. Singh, J.K. Jansson, A. Probst, S.E. Borglin, J.L. Fortney, W.T. Stringfellow, M. Bill, M.E. Conrad, L.M. Tom, K.L. Chavarria, T.R. Alusi, R. Lamendella, D.C. Joyner, C. Spier, J. Baelum, M. Auer, M.L. Zemla, R. Chakraborty, E.L. Sonnenthal, P. D'Haeseleer, H.-Y.N. Holman, S. Osman, Z. Lu, J.D. Van Nostrand, Y. Deng, J. Zhou and O.U. Mason. 2010. Deep-sea oil plume enriches indigenous oil-degrading bacteria. *Science*, 330, pp. 204–208.

Johansen, Ø., P.J. Brandvik and U. Farooq. 2013. Droplet breakup in subsea oil releases–Part 2: Predictions of droplet size distributions with and without injection of chemical dispersants. *Marine Pollution Bulletin*, 73(1), pp. 327–335.

Johansen, Ø., M. Reed and N.R. Bodsberg. 2015. Natural dispersion revisited. *Marine pollution bulletin*, 93(1–2), pp. 20–26.

Jones, C.E., K.F. Dagestad, Ø. Breivik, B. Holt, J. Röhrs, K.H. Christensen, M. Espeseth, C. Brekke and S. Skrunes. 2016. Measurement and modelling of oil slick transport. *Journal of Geophysical Research: Oceans*, 121(10), pp. 7759–7775.

Kotzakoulakis, K. 2020a. Subsea oil-well blowout, near-field plume model, created on May 11, 2020 in Youtube. https://youtu.be/KHQUztxgG50

Kotzakoulakis, K. 2020b. Subsea oil-well blowout, coupled near-field & far-field model (A), created on May 10, 2020 in Youtube. https://youtu.be/SYQInZARwRY

Kotzakoulakis, K. 2020c. Subsea oil-well blowout, coupled near-field & far-field model (B), created on May 10, 2020 in Youtube. https://youtu.be/8bGgHOEFoO4

Kotzakoulakis, K. 2020d. Effect of droplet size on far-field oil-spill circulation, created on May 10, 2020 in Youtube. https://youtu.be/Jk1MBaTbuP0

Kotzakoulakis, K., S.N. Estrada-Allis, A. Maslo and J. Sheinbaum. 2019. Advancements in the capabilities of the OpenDrift model for the prediction of the fate and the circulation of oil spills in the Gulf of Mexico. *Reunión Anual de la Unión Geofísica Mexicana (RAUGM) 2019*. Puerto Vallarta, Jalisco, Mexico.

Kotzakoulakis, K. and S.C. George. 2018. Predicting the weathering of fuel and oil spills: A diffusion-limited evaporation model. *Chemosphere*, *190*, pp. 442–453.

Li, Z., M.L. Spaulding andD. French-McCay. 2017. An algorithm for modeling entrainment and naturally and chemically dispersed oil droplet size distribution under surface breaking wave conditions. *Marine pollution bulletin*, *119*(1), pp. 145–152.

Malone, K., S. Pesch, M. Schlüter and D. Krause. 2018. Oil droplet size distributions in deep-sea blowouts: Influence of pressure and dissolved gases. *Environmental Science & Technology*, *52*(11), pp. 6326–6333.

McNutt, M.K., R. Camilli, T.J. Crone, G.D. Guthrie, P.A. Hsieh, T.B. Ryerson, O. Savas and F. Shaffer. 2012. Review of flow rate estimates of the Deepwater Horizon oil spill. *Proceedings of the National Academy of Sciences*, 109(50), pp. 20260–20267.

Nissanka, I.D. and P.D. Yapa. 2017. Oil slicks on water surface: Breakup, coalescence, and droplet formation under breaking waves. *Marine Pollution Bulletin*, *114*(1), pp. 480–493.

NOAA PyGnome model repository at GitHub. https://github.com/NOAA-ORR-ERD/PyGnome

North, E.W., E.E. Adams, A.E. Thessen, Z. Schlag, R. He, S.A. Socolofsky, S.M. Masutani and S.D. Peckham. 2015. The influence of droplet size and biodegradation on the transport of subsurface oil droplets during the Deepwater Horizon spill: A model sensitivity study. *Environmental Research Letters*, 10(2), p. 024016.

Ohnesorge, W.V. 1936. Die bildung von tropfen an düsen und die auflösung flüssiger strahlen. *ZAMM-Journal of Applied Mathematics and Mechanics/Zeitschrift für Angewandte Mathematik und Mechanik*, *16*(6), pp. 355–358.

OpenDrift-TAMOC model repository at GitHub. https://github.com/KKotzak/OpenDrift-TAMOC

Paiva, P.M., J.L. Junior, A.N. Barreto, J.A.F. Silva and A.S. Neto. 2017. Comparing 3d and 2d computational modelling of an oil well blowout using MOHID platform-A case study in the Campos Basin. *Science of the Total Environment, 595*, pp. 633–641.

RPS Oilmap Deep. *Oxfordshire*. United Kingdom. https://www.rpsgroup.com/services/oceans-and-coastal/modelling/products/oilmapdeep/

RPS SIMAP. *Oxfordshire*. United Kingdom. http://asascience.com/software/simap/

Schwartzberg, H.G. 1971. The movement of the oil spills. *International Oil Spill Conference Proceedings, 1971*, pp. 489–494, https://doi.org/10.7901/2169-3358-1971-1-489.

Sim, L., J. Graham, K. Rose, R. Duran, J. Nelson, J. Umhoefer and J. Vielma. 2015. *Developing a comprehensive deepwater blowout and spill model*. NETL-TRS-9-2015; EPAct Technical Report Series, U.S. Department of Energy, National Energy Technology Laboratory, Albany, OR, p 44. doi:10.18141/1432302.

SINTEF OSCAR DeepBlow. *Trondheim*. Norway. https://www.sintef.no/en/software/oscar/

Socolofsky, S.A., E.E. Adams, M.C. Boufadel, Z.M. Aman, Ø. Johansen, W.J. Konkel, D. Lindo, M.N. Madsen, E.W. North, C.B. Paris and D. Rasmussen. 2015b. Intercomparison of oil spill prediction models for accidental blowout scenarios with and without subsea chemical dispersant injection. *Marine Pollution Bulletin, 96*(1–2), pp. 110–126.

Socolofsky, S.A., A.L. Dissanayake, I. Jun, J. Gros, J.S. Arey and C.M. Reddy. 2015a. Texas A&M oil spill calculator (TAMOC): Modelling suite for subsea spills. In *Proceedings of the thirty-eighth AMOP technical seminar*. Environment Canada Ottawa, pp. 153–168.

The climate data guide: climate forecast system reanalysis (CFSR). https://climatedataguide.ucar.edu/climate-data/climate-forecast-system-reanalysis-cfsr. Last modified 08 Nov 2017.

Tkalich, P. and E.S. Chan. 2002. Vertical mixing of oil droplets by breaking waves. *Marine Pollution Bulletin, 44*(11), pp. 1219–1229.

Vitu, S., J.N. Jaubert and F. Mutelet. 2006. Extension of the PPR78 model (Predictive 1978, Peng–Robinson EOS with temperature dependent kij calculated through a group contribution method) to systems containing naphthenic compounds. *Fluid Phase Equilibria, 243*(1–2), pp. 9–28.

Wang, C.Y. and R.V. Calabrese. 1986. Drop breakup in turbulent stirred-tank contactors. Part II: Relative influence of viscosity and interfacial tension. *AIChE journal, 32*(4), pp. 667–676.

Xu, X., W. Liu, S. Tian, W. Wang, Q. Qi, P. Jiang, X. Gao, F. Li, H. Li and H. Yu. 2018. Petroleum hydrocarbon-degrading bacteria for the remediation of oil pollution under aerobic conditions: A perspective analysis. *Frontiers in microbiology, 9*, p. 2885.

Zhao, L., Boufadel, M.C., Socolofsky, S.A., Adams, E., King, T. and Lee, K., 2014. Evolution of droplets in subsea oil and gas blowouts: Development and validation of the numerical model VDROP-J. *Marine Pollution Bulletin, 83*(1), pp. 58–69.

Zhao, L., F. Shaffer, B. Robinson, T. King, C. D'Ambrose, Z. Pan, F. Gao, R.S. Miller, R.N. Conmy and M.C. Boufadel. 2016. Underwater oil jet: Hydrodynamics and droplet size distribution. *Chemical Engineering Journal, 299*, pp. 292–303.

Zheng, L. and P.D. Yapa. 2000. Buoyant velocity of spherical and nonspherical bubbles/droplets. *Journal of Hydraulic Engineering, 126*(11), pp. 852–854.

9 Economic and Financial Impacts of Oil Spills

Liangliang Lu and Pentti Kujala

Aalto University

School of Engineering

Department of Mechanical Engineering

Marine Technology

This chapter describes the impacts of oil spills from the economic and financial perspectives. Before entering the topic of oil spill impacts, the definitions of the economics and finance will be introduced first so that the scope and perspective of analyzing the oil spill impact are clarified.

Economics is understood as a social science which studies the production, consumption, and distribution of goods and services, with the aim of explaining how economies work and how their agents interact. Modern economics is in fact often very quantitative and heavily math-oriented in practice and divided into two branches: macroeconomics and microeconomics. Macroeconomics studies how the aggregate economy behaves, such as inflation, national income, gross domestic product (GDP), and changes in unemployment. Microeconomics is the study of economic tendencies, focusing on the smaller factors that affect choices made by individuals and companies.

Finance in many respects is an offshoot of economics and can be divided into three categories: public finance, corporate finance, and personal finance. It describes the management, creation, and study of money, banking, credit, investments, assets, and liabilities that make up financial systems, as well as the study of those financial instruments.

Based on the understanding of the abovedescribed definitions of economics and finance, the analysis of the oil spill impact from an economic and financial perspective then needs to examine the oil and oil spill system, which interact with the society. Therefore, the following section will first introduce the oil and oil spill response and recovery-related system in Section 9.1 and the economic and financial impact of oil spill will then be introduced in Section 9.2, based on the identified system in Section 9.1. Major oil spill examples will be discussed in Section 9.3.

9.1 OIL, OIL SPILL, RESPONSE AND RECOVERY SYSTEM

This section aims to introduce how and where the oil spills may happen and what the corresponding response and recovery options are. First, the oil and gas system will be described so that we can know what the potential oil spill scenarios are and where they may happen. Second, the oil response and recovery system will be introduced briefly.

9.1.1 Oil and Gas System and Potential Oil Spills

Knowing the oil and gas industry system helps to understand where the oil spills may happen and their impacts inside and outside the system, which will facilitate the understanding of economic and fanatical aspects.

9.1.1.1 Oil and Gas System

The oil and gas industry system is usually categorized into three sectors: upstream, midstream, and downstream (Devold, 2013), as illustrated in Figure 9.1.

The upstream sector is also known as the exploration and production (E&P) sector. It consists of processes and operations that involve searching for potential underground or underwater crude oil and natural gas fields, drilling of exploratory wells, and subsequently drilling and operating the wells that recover and bring the crude oil and/or raw natural gas to the surface. This sector can be onshore or offshore. Offshore exploration and production involve exploration ships and different offshore platforms or floating production storage and offloading (FPSO) ships, floating liquefied natural gas (FLNG) ships, etc., and relevant facilities.

The midstream sector includes the transport and storage of crude oil before refining and processing it into fuels and elements that have become indispensable in our day-to-day life. Transportation options can vary from small connector pipelines to massive cargo ships making trans-ocean crossings, depending on the commodity and distance covered. After being driven to the plant it is stored in large tanks.

The refining, commercialization, and distribution of products derived from crude oil and natural gas constitute the downstream sector. In the refineries, the crude undergoes a process of physical separation and then chemical processes to extract

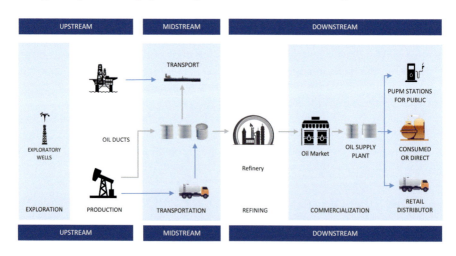

FIGURE 9.1 Oil and gas system diagram.

Economic and Financial Impacts of Oil Spills 267

the different components that form it. The extracted derivates are fuels and petrochemicals. Thus, consumers receive products such as gasoline, kerosene, diesel, lubricants, asphalt, natural gas and liquefied gas, synthetic rubber, fertilizers, preservatives, packaging, and plastics. Wholesale and retail marketing activities help move the finished products from energy companies to retailers or end users.

9.1.1.2 Potential Oil Spills

Understanding the oil system helps in seeing potential oil spills which may happen during the whole process. Table 9.1 lists the potential oil spill sources for different stages for onshore and offshore systems.

In the upstream stage, oil may be spilled from wells and pipelines during onshore and offshore activities. In addition, different from onshore oil spill, offshore oil spill sources also include platforms and ships (e.g. FPSO or FLNG). In midstream, the differences lie in the transportation. In addition to pipelines, ships are necessary to transport the drilled oil from offshore to shore. Onshore oils are supposed to be transported by pipelines or onshore transportations. However, due to the global features of oil, oil-refining facilities are usually located in different places. For example, the main onshore oil well may be located in the Middle East, while the oil-refining factories are in Europe and Asia. Therefore, even the onshore oils need ships for transportation, which makes little difference for onshore and offshore potential oil spill sources. Downstream, oil spill sources onshore and offshore have almost no difference. After refining, oils will be distributed to different industries and customers. It can be inside cars, buses, and oil gas stations in cities. It can also be in airplanes or ships as fuel. The ships will be operated then for different purposes, including the operations in upstream and midstream. Meanwhile, there are some other sources, e.g. wrecks on the seabed which still contain oils. In addition, offshore oil spills may

TABLE 9.1
Potential Oil Spill Sources

	Onshore	Offshore
Upstream	Well	Well
	Pipeline	Pipeline
		Platform
		Ship
Midstream	Pipeline	Pipeline
	Truck/train/ship	Ship/truck/train
	Storage	Storage
Downstream	Refiners	Refiners
	Oil and gas station	Oil and gas station
	Airplane	Airplane
	Ship	Ship
	Car/bus/truck	Car/bus/truck
Other		Wreck

268   Oil Spill Occurrence, Simulation, and Behavior

also happen in sea ice conditions, which gives more complex situations (Lu et al., 2019; Lu et al., 2020).

From the potential oil spill sources, the complexity and severity of oil spill can be assessed somehow. It can be seen that offshore oil spills are usually more complex as the environment is more complicated and not easy to control. Meanwhile upstream usually has larger possibilities for bigger oil spills. Although oil spills in refineries and storage can also be large, the monitoring system and response system can be timely so that the consequences and severity are relatively easy to control. In contrast, offshore oil spills are usually difficult to control, and it also takes a long time for response facilities to reach the spill and conduct measures. Therefore, the focuses in this chapter are on the offshore oil spill cases, including upstream sources and the ships in other stages. Statistically, it can also be seen that major oil spill accidents are from offshore activities; see more examples in Section 9.3.

9.1.2 OIL SPILL RESPONSE AND RECOVERY SYSTEM

This section aims to briefly describe the oil spill response and recovery system so that the relevant elements can clearly stand out in this period, and this will be beneficial in communicating the economic and financial parts.

A tiered approach (IMO, 2010; EPPR, 2015) has been developed and adopted for different oil spills. The approach has three tiers and can be described as follows:

> Tier 1: operational-type spills that may occur at or near a company's own facilities, as a consequence of its own activities. An individual company would typically provide resources to respond to this type of spill. Commercial ships are required to have a shipboard oil pollution emergency plan (SOPEP) to deal with Tier 1 level spills. Details of plan requirements vary with the nation involved.
> Tier 2: a larger spill in the vicinity of a company's facilities where resources from other companies, industries, and possibly government response agencies in the area can be mobilized on a mutual aid basis. The company may participate in a local co-operative where each member pools their Tier 1 resources and has access to equipment that may have been jointly purchased by a co-operative.
> Tier 3: a large spill where substantial further resources may be required and support from a national or international co-operative stockpile may be necessary. It is likely that such operations would be subject to government controls and direction through a combined inter-agency/industry command center.

An good response and recovery measures is to select and implement a combination of response techniques that would be most effective in minimizing the overall short- and long-term impact, including preventing oil from reaching the shoreline and sensitive areas. The response countermeasures generally include three types: (1) mechanical containment and recovery by utilizing booms and skimmers; (2) a

Economic and Financial Impacts of Oil Spills

combination of strategies to concentrate the oil and burn in-situ; (3) dispersants that disperse surface oil into the water column as small oil droplets for microorganisms to biodegrade effectively. All the response options need ships to deploy booms to stop further oil spreading. In addition, relevant response ships equipped with inside skimmer systems or mounted external skimmers are needed to conduct the recovery process if the mechanical recovery approach is adopted. The in-situ burning and dispersants approaches require environmental favor in the conditions, e.g. suitable oil film thickness, waves, winds, degree of emulsification, etc. The aerial application of proven herding agents and ignitors is another possible rapid tool. Dispersants are usually distributed by fast aircraft for large spills to improve the response time. In addition to the response countermeasures, detection, monitoring, and tracking are important steps among the response actions. It is crucial to know where the spilled oil is and the condition of that oil (degree of weathering) at any given time. Therefore, a mix of airborne sensors, remote sensing, including marine or satellite radar systems, and even trajectory analysis and transportation modeling may be applied. Unmanned aerial systems (UAS) and autonomous underwater vehicles (AUVs) can also be helpful in some cases. When the spilled oil comes from the seabed well or subsea risers, remote operation vehicles (ROVs) and other oil recovery equipment, e.g. containment domes, will also be used to control or stop the oil spill. In extreme cases, other drilling platforms may be required to make additional drilling to permanently block the wells. Before the oil reaches the shoreline, protective booms are usually used to protect the coastal areas from the oil. After the oil reaches the shoreline, shoreline oil recovery or removal becomes the priority, using machines and human labor.

9.2 ECONOMIC AND FINANCIAL ASPECTS OF OIL SPILL

After describing the oil, oil spill, and response and recovery system, the focus in this section moves to the economic and financial aspects of an oil spill. The emphasis is still on offshore oil spill cases.

Figure 9.2 shows a diagram of an oil spill impact framework (Chang et al., 2014). The impact of the oil spill can be described from short-term and long-term perspectives, which link two main parts: ecological and societal impacts. When an oil spill happens, the spilled oil will be exposed to the external environment, and the transportation and weathering start to happen with the interaction with the environmental conditions. Meanwhile, the corresponding response actions usually also begin to remove the oil and mitigate the oil spill impacts. Thus, the first economic impact is from the response actions which include various cooperation, equipment, and resource deployments. The further weathered oil then begins to influence the ecosystems in the places to which it is transported. The regional ecosystem species, mobility, and habitat will be affected and even cause damage in the long term. This leads to natural resource damage. In addition, the impacted ecosystem will also influence the society and its economy through fisheries, tourism and recreation, etc. The socioeconomic impacts will interact regionally and even globally through the global economic and financial system and cause long-term impacts. There are also social impacts on human beings, e.g. health and physiological impacts, and there will

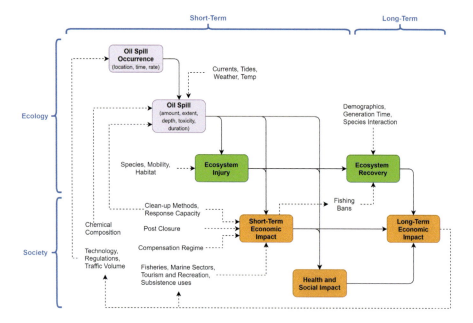

FIGURE 9.2 Oil spill impact framework. (Boxes = outcomes, lowercase = variables, solid lines = linkages between oil spill occurrence and socioeconomic impacts, dotted lines = linkages between exogenous variables and outcomes) (after Chang et al., 2014).

also be more research activities going on after the oil spills. All of these will have a further impact on the technology, regulations, and marine activities, e.g. traffic volume, oil drilling, etc.

Based on the diagram of how an oil spill influences the ecology and society in Figure 9.2, the following sections are going to summarize an overall structure of the economic and financial impacts from an oil spill, as shown in Figure 9.3.

First, the economic impact is divided into local and global impacts. Local impacts refer to the damage or costs to the regional environment and society. They include cleanup costs, socioeconomic loss, natural resource damage, and other costs. In the global aspect, the impact is transferred through the global financial market and has an influence on the global society. The impact may be reflected in the oil price, stock market, industry, and regulations, which further influence the global economy

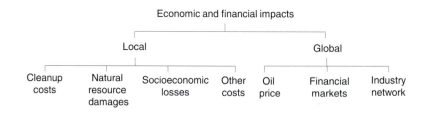

FIGURE 9.3 Economic impacts from oil spills.

Economic and Financial Impacts of Oil Spills **271**

and finance. This global impact exists mainly for the major oil spills. The local and global impacts will be discussed in the following sections in detail.

9.2.1 Local Economic Impact

The local economic impact refers to the oil spill-induced impacts related to regional economic damage and loss, including cleanup costs, natural resource damage, socioeconomic loss, and other costs in the oil-spilled water and region. The loss and damages will influence individuals and companies microscopically and are also reflected in the macroeconomy including in inflation, regional or national income, gross domestic product (GDP), and changes in unemployment, etc. It is then a very complicated system to study. Therefore, the following sections will restrict the target to the more straightforward impacts: the cleanup costs, natural resource damage, socioeconomic loss, and other corresponding costs caused by oil spills, with a focus on the estimation procedures and methods.

9.2.1.1 Cleanup Costs

The cleanup of oil covers the removal of oil from the sea, coastal waters, and shorelines as well as the disposal of the collected oil waste. It includes the command center coordinating cost, offshore oil treatment (mechanical recovery, dispersant spraying, or in-situ burning), nearshore recovery and protection, shoreline and wildlife cleanup, etc. Due to the complexity of the cleanup procedure and wide involvement of the various resources, there are several approaches to estimate the cleanup cost for an oil spill accident.

9.2.1.1.1 Estimation Model Based on Causation

The first one is the estimation based on the causal cost components, more like direct estimation, which calculates the costs by adding each cost from various cost sources. The general equation can be expressed as:

$$C_c = \sum_{t=1}^{n} C_t \tag{9.1}$$

Where C_c is the total cleanup cost, C_t is the cost used for different type of resources, t means the resource category, and n is the total number of the categories.

The general equation is simple; however more calculation and estimation efforts are needed for digging into the details in each cost category. The next sections will show two sample methods of the cleanup cost estimation.

9.2.1.1.1.1 Oil Spill Cost Study in UK A study on oil spills (Oil & Gas UK and OPOL, 2012) was conducted to estimate the oil spill-induced costs. Four spill scenarios were studied, as shown in Table 9.2. Oil transportation and weathering were firstly modeled by running the simulation model, and then the oil distribution can be understood to further estimate the induced cleanup cost.

TABLE 9.2
Oil Spill Scenarios in the Case Study

Location	Lat/Lon	Oil Type	Release Duration (Days)	Release Rate (10^3 Barrels per Day)	Model Run Duration (Days)	ITOPF Group Classification
1	60° 25'N/ 4° 6' W	Schiehallion	10	31	10	G3
2	58° 5' N/ 1°42' W	Captain	10	18	10	G3
3	61° 35' N/ 1° 18' E	Magnus	10	58	10	G2
4	57° 45' N/ 0° 54' E	Forties	10	41	10	G2

Source: Oil & Gas UK and OPOL (2012).

Direct cost estimation was applied by setting nine cost categories: (1) command center costs, (2) offshore dispersant spraying, (3) offshore mechanical recovery, (4) nearshore mechanical recovery, (5) protective nearshore booming, (6) shoreline cleanup, (7) wildlife cleanup costs, (8) shoreline cleanup assessment team (SCAT), media liaison, and surveillance, (9) disposal costs.

In each category, there are variable-per-day costs and fixed-variable costs and all the detailed costs for each item are listed in each category, which allows the direct calculation of the cleanup cost by the following equation:

$$C_t = \sum_{i,j=1}^{p,q} C_{Vi} N_{Vi} T_{Vi} + C_{Fj} N_{Fj} \tag{9.2}$$

Where C_t is the cleanup cost for category t, C_{Vi} is the cost per day for variable i in category t, N_{Vi} is required amount for variable i per day, and T_{Vi} is the estimated days needed for conducting variable i activity. C_{Fj} is the fixed cost per unit for variable j, N_{Vj} is required amount for variable j. p and q are the total numbers of the variables per day and fixed variables respectively in each category.

The method here is simple, and it can be calculated inside MS Excel by setting the equations. However, the emphasis in this example is on the identified categories and the detailed cost information about each resource in the categories. Getting the right variable and cost information becomes essential in this kind of direct estimation. More information can be found in the report (Oil & Gas UK and OPOL, 2012). It should be noted that the cost information can be just used for reference as the costs

Economic and Financial Impacts of Oil Spills

TABLE 9.3

Estimated Cleanup Costs in the Case Study (Unit: Million $)

Remedial measures	Location 1		Location 2		Location 3		Location 4	
	Low	High	Low	High	Low	High	Low	High
Shoreline cleanup	25.6	88.0	30.4	104.0	5.1	17.6	1.8	5.9
Offshore mechanical recovery	20.8	20.8	20.8	20.8	20.8	20.8	20.8	20.8
Protective nearshore booming	24.0	24.0	22.4	22.4	30.4	30.4	30.4	30.4
Command center	13.9	13.9	20.8	20.8	30.4	30.4	30.4	30.4
Media and SCAT	8.5	8.5	8.5	8.5	8.5	8.5	8.5	8.4
Wildlife response	2.9	2.9	2.9	2.9	2.9	2.9	2.9	2.9
Disposal	2.1	2.1	2.1	2.1	3.2	3.2	3.2	3.2
Total	13.9	67.2	15.8	43.2	15.0	19.2	11.2	11.2

Source: Oil & Gas UK and OPOL (2012).

vary in different locations and different scenarios. Table 9.3 gives the estimated costs in the case study.

9.2.1.1.1.2 Cleanup Cost Study in Finland Other studies on the oil spill cleanup cost estimation were carried out in Finland (Montewka et al., 2013; Helle et al., 2015), which belong to the category of estimation based on causal cost components. The principle of the studies is adding up all costs for different resources. However, the difference is that the Bayesian network method is applied, which leads to a probabilistic model. As shown in Figure 9.4, cleanup costs are divided into offshore and onshore and further broken down to detailed categories. The offshore costs include preparation costs, booms, combating cost, the cost of emptying tanks, and air surveillance cost. The onshore cleanup costs cover machine cost, boat cost, and manual cost. All the sub-costs, except booms cost, rely on the predicted time through the probabilistic network. The per unit time cost information of different resources is based on the literature, and expert opinions. Table 9.4 shows the cleanup cost estimations in the case studies.

With a similar Bayesian network model approach, Helle et al. (2015) separated the costs further into combating operations and waste-treatment operations for both open-sea and shoreline responses and recoveries. The calculated costs are summarized in Table 9.5 for the case studies.

9.2.1.1.2 Estimation Model Based on Oil Spill Database

This type of estimation for cleanup cost is more based on large data from historical oil spill accidents. Three models are introduced in the following sections: (1) the Etkin model, (2) a limit-parameter-based regression model, and (3) a regression model based on compensation data.

274   Oil Spill Occurrence, Simulation, and Behavior

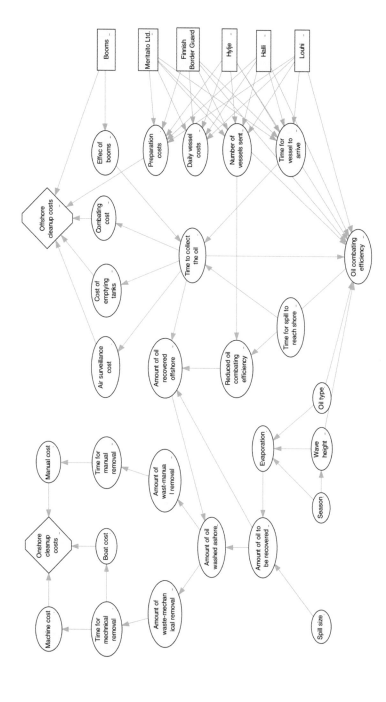

FIGURE 9.4 Cleanup cost estimation BN model (after Montewka et al., 2013).

Economic and Financial Impacts of Oil Spills

TABLE 9.4

Estimated Cleanup Cost (Unit: Million €)

Scenario	1	2
Spill volume (m³)	5000	50,000
Oil type	Medium oil	Heavy crude oil
Cleanup cost	13.5	98.1

Source: Montewka et al. (2013).

TABLE 9.5

Estimated Cleanup Cost (Unit: Million €)

Scenario	1	2
Spill volume (m³)	3000	50,000
Oil type	Medium oil	Heavy crude oil
Combating operation		
Open sea	1.5	1.5
Shoreline	4	14.7
Waste treatment operation		
Open sea	0.5	3.7
Shoreline	2.7	29
Total	8.7	49

Source: Helle et al. (2015).

9.2.1.1.2.1 Etkin Model

9.2.1.1.2.1.1 General Cleanup Cost Etkin (1999, 2000) developed a method for estimating cleanup costs (per unit ton or gallon). The approach is based on historical data over more than 200 oil spills worldwide. The analysis focuses on the per-unit (per-ton or per-gallon) oil spill cleanup costs of various factors. The estimation formulas were derived as:

$$C_{ei} = C_{ui}A_i \tag{9.3}$$

$$C_{ui} = C_{li}t_io_im_is_i \tag{9.4}$$

$$C_{li} = C_n r_i l_i \tag{9.5}$$

where C_{ei} is the estimated total response cost for the scenario, i; A_i is the specified spill amount for the scenario; C_{ui} is the response cost per unit for scenario, i; C_{li} is the cost per unit spilled for scenario, i; C_n is the general cost per unit spilled in a country, n, as shown in Table 9.6; t_i is the oil type modifier factor for scenario, i; o_i is the

TABLE 9.6

General Cost per Unit Spilled in Countries

Country	USD/Gallon	USD/Ton
North America		
Canada	22.1	6508.1
United States	87.1	25,614.6
Average	67.4	19,814.6
Latin America		
Argentina	7.9	2316.6
Brazil	19.0	5600.7
Chile	3.1	910.4
Mexico	2.9	850.3
St. Kitts/Nevis	10.5	3085.8
Uruguay	11.5	3368.3
Venezuela	40.2	1817.8
Average	10.4	3055.8
Africa		
Egypt	15.1	4428.9
Morocco	32.9	9675.1
Mozambique	0.04	6.1
Nigeria	6.0	1766.8
South Africa	9.9	2917.5
Average	10.7	3163.9
Europe		
Denmark	38.0	11,180.4
Estonia	23.2	6820.6
Finland	7.2	2115.3
France	7.8	2301.6
Germany	36.4	10,702.7
Greece	29.0	8530.3
Ireland	16.4	4807.5
Italy	22.3	6541.2
Latvia	31.3	9212.4
Lithuania	0.26	78.1
Netherlands	22.6	6655.4
Norway	78.6	23,118.1
Spain	1.5	438.7
Sweden	53.2	15,642.4
United Kingdom	10.5	3082.8
Yugoslavia	5.2	1541.4
Average	36.7	10,807.8
South Pacific		
Australia	20.4	5991.3

(Continued)

Economic and Financial Impacts of Oil Spills

TABLE 9.6 (CONTINUED)
General Cost per Unit Spilled in Countries

Country	USD/Gallon	USD/Ton
New Zealand	9.5	2791.4
Average	19.4	5698.9
Middle East		
Israel	7.9	2313.6
UAE	2.2	637.0
Average	3.6	1057.5
Asia		
Hong Kong, China	15.1	4452.9
Japan	117.8	34,619.9
Malaysia	260.9	76,589.3
Philippines	2.3	676.5
Singapore	1.3	390.6
South Korea	43.6	12,814.9
Average	93.5	27,495.8

Source: Etkin (2000).

shoreline oiling modifier factor for scenario, i; m_i is the cleanup methodology factor for scenario, i; s_i is the spill size modifier factor for scenario, i; r_i is the regional location modifier factor for scenario, i; and l_i is the local location modifier for scenario, i. The modifiers are listed in Table 9.7.

9.2.1.1.2.1.2 On-Water Cleanup Cost   A further calculation tool was developed by Etkin and Welch (2005), which further divides costs into on-water and shoreline response costs based on case studies and the modeling of hypothetical spills. The estimation formulas for on-water response cost can be summarized as:

$$C_{rw} = C_{rw1} + C_{rw2} + C_{rw3} + C_{rw4} \tag{9.6}$$

$$C_{rw1} = 3236.7A^{0.4387} \tag{9.7}$$

$$C_{rw2} = 35388A^{0.2762} \tag{9.8}$$

$$C_{rw3} = C_{uwi}A \tag{9.9}$$

$$C_{rw4} = 5P_r P_w A \tag{9.10}$$

Where C_{rw} is the on-water response cost, C_{rw1} is the monitoring cost, C_{rw2} is the mobilization cost, C_{rw3} is the oil-related cost, and C_{rw4} is the disposal cost. C_{uwi} is the cost per gallon by oil type and response type, as shown in Table 9.8. A is spill volume, P_r is the percent of oil remaining, and P_w is the percent of on-water removal.

TABLE 9.7
Cost Factors for Modifiers

Cost Factor	Modifier
Oil Type	
No. 2 fuel (diesel)	0.18
Light crude	0.32
No. 4 fuel, No. 5 fuel	1.82
Crude	0.55
Heavy crude	0.65
No. 6 fuel	0.71
Spill Size (gal/ton)	
0–1E04/0–34	2.00
1E04–1E05/34–340	0.65
1E05–5E05/340–1700	0.27
5E05–1E06/1700–3400	0.15
1E06–1E07/3400–34,000	0.05
1E07+/34,000+	0.01
Spill Location	
Nearshore	1.46
In port	1.28
Offshore	0.46
Primary Cleanup Method	
Dispersants	0.46
In-situ burning	0.25
Mechanical	0.92
Manual	1.89
Natural cleansing	0.10
Shoreline (km)	
0.1	0.47
2.2	0.54
20.9	0.61
100	1.06
500	1.53

Source: Etkin (2000).

9.2.1.1.2.1.3 Shoreline Cleanup Cost   In addition to the on-water cleanup cost, Etkin and Welch (2005) described the corresponding calculations for shoreline response cost. The estimation formula is concluded here as:

$$C_{sr} = f_{HCIC} \sum_{i=1}^{n} C_{usi} S_i f_{ci} \qquad (9.11)$$

Economic and Financial Impacts of Oil Spills

TABLE 9.8
Per Gallon Cost (USD) by Oil Type and Response Type

Oil Type	Response Type		
	Mechanical	Burn	Dispersant
Medium crude	28	23	24
Gasoline	12	NA	NA
Jet fuel	12	NA	NA
Diesel	15	13	14
Light crude	23	18	19
Heavy crude	36	29	30
No. 6 fuel	37	30	31
IFO	28	28	24
Lube oil	26	21	22

NA means the response type is not recommended for that oil type, thus no cost data.
Source: Etkin and Welch (2005).

Where C_{sr} is the shoreline response cost, C_{usi} is the shoreline response cost per area by oil thickness for shoreline type i, S_i is the shoreline oiled area, f_{ci} is the coverage adjustment factor, and f_{HCIC} is the How Clean Is Clean (HCIC) adjustment factor. Table 9.9 shows an example of C_{usi} for medium crude oil type.

TABLE 9.9
Medium Crude Oil Shoreline Spill Cost (USD/m²)

Shoreline Type	Pooled Oil 1+ cm	Cover 0.1–1 cm	Coat 0–0.1 cm	Stain	Film
1A: Exposed rocky	159	59	15	6	3
2: Rocky platform	159	59	15	6	3
3: Fine sand	159	59	15	6	3
4: Coarse sand	159	59	15	6	3
5: Mixed sand/gravel	193	71	18	8	4
6A: Gravel beach	193	71	18	8	4
6B: Riprap structures	193	71	18	8	4
7: Exposed tidal flat	259	96	24	10	5
8A: Sheltered rocky shore	159	59	15	6	3
8B: Sheltered solid	125	46	12	5	3
9: Sheltered tidal flat	259	96	24	10	5
10A: Salt/brackish marsh	259	96	24	10	5
10B: Freshwater marsh	259	96	24	10	5
10C: Swamp	259	96	24	10	5
10D: Mangrove	259	96	24	10	5

Source: Etkin and Welch (2005).

$$f_c = \begin{cases} 3.0, \, C, \text{continous} \\ 2.3, \, B, \text{broken} \\ 1.0, \, P, \text{patchy} \\ 0.2, \, S, \text{sporadic} \end{cases} \tag{9.12}$$

$$f_{HCIC} = \begin{cases} 2.7, \text{Maximum} \\ 1.2, \text{High} \\ 1.0, \text{Medium} \\ 0.8, \text{Low} \\ 0.4, \text{Minimum} \end{cases} \tag{9.13}$$

9.2.1.1.2.2 A Limit-Parameter-Based Regression Model Shahriari and Frost (2008) developed a mathematical method to estimate cleanup costs based on regression analysis of 80 incidents during the period of 1967–2002. The model parameters are spill quantity, oil density, distance to shore, cloudiness, and level of preparedness. In order to reach predictions as reliable as possible, the composite model is used, i.e.

$$C_{c1} = 156.5934A + 5678100\rho + 2303500L_p - 4979000 \tag{9.14}$$

$$C_{c2} = \left(29471\rho + 863.0906L_p - 24060\right)A \tag{9.15}$$

Where C_{c1} and C_{c2} are cleanup cost estimations by two approaches, A is the spill amount in tons, ρ is oil density in kg/m³, L_p is level of preparedness, on a scale of 1–3, 3 being best.

The model user needs to use both equations and then decide which prediction to use, based on the cost interval [4×10^6 to 4×10^7]. If both predictions end up within the interval, choose C_{c1}. If both end up outside the interval, choose C_{c2}.

9.2.1.1.2.3 A Regression Model Based on Compensation Data Kontovas et al. (2010) developed a method to estimate the cleanup cost by using the data of compensations paid by the International Oil Pollution Compensation Funds (IOPCF). The formula developed is expressed as:

$$C = 44435A^{0.644} \tag{9.16}$$

Where C is cleanup cost in US$ and A is spill size in tons.

9.2.1.2 Natural Resource Damages

9.2.1.2.1 Estimation Based on Recovery Function

A degradation of natural resources will happen after an oil spill and, consequently, decrease their services. Liu and Wirtz (2009) developed an estimation method for

Economic and Financial Impacts of Oil Spills

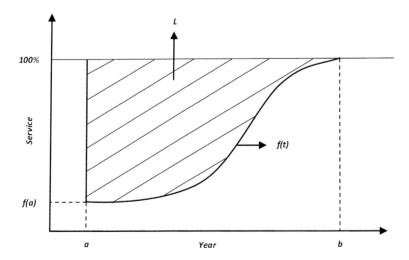

FIGURE 9.5 Time integral over lost service L. f(t) is a recovery function to describe the potential service provided by the injured habitats during the year t.

lost service value of natural resources after oil impacts. As shown in Figure 9.5, the lost services are represented by the time-integrated area L, which can be estimated in terms of the fraction f(t) of intact services. The fraction f(t) can be a time-dependent recovery function; then L sums over the years of loss until the injured resource is completely restored. The formula can be expressed as:

$$L = \sum_{t=a}^{b} \left(1 - f(t)\right)\left(\frac{1}{1+d}\right)^{t-a} \tag{9.17}$$

Where d is the yearly discount rate reflecting that the present service losses are more costly than the future ones. A value of $d=0.03$ is recommended (NOAA, 1999).

The total value lost, T, is expressed as an aggregation over individual losses related to each resource:

$$T = \sum_{i=1}^{n} V_i = \sum_{i=1}^{n} M_i Q_i L_i \tag{9.18}$$

Where Q_i is the total amount of injured resource i and M_i is monetary value per unit for resource i, and n is the number of resources damaged by the released oil.

M_i is difficult to define, as most of the environmental goods and services are non-marketed. Currently, there are two key ways to get the unit value of non-market resources: (1) measure through direct or indirect economic means, (2) transfer an estimated value from past studies (Liu and Wirtz, 2006). The first way requires application of valuation methods to measure the value, and there are four categories (Turner et al., 1995): the constructed market valuation (CMV) methods, surrogate market valuation (SMV) methods, market-oriented valuation (MOV) methods, and

282 Oil Spill Occurrence, Simulation, and Behavior

cost-based valuation methods (CBV). The details of the methods will not be discussed here. It should be noted that the value of environmental resources discussed here refers to both use and non-use values. The estimates derived from the CMV, SMV, and MOV methods emphasize losses of value based on the people's willingness to pay (WTP) in a real or hypothetical market.

9.2.1.2.2 Estimation Model Based on Oil Spill Database

Etkin and Welch (2005) also include an estimation for natural resource damage. The formulas for on-water natural resource damage are summarized here as:

$$T = \sum_{i=1}^{n} V_i = \sum_{i=1}^{n} S_i G d_o G^{-x_o} \tag{9.19}$$

Where T is the total value lost, V_i is the lost value for natural resource i, S_i is the natural resource i modifier, G is the spill amount in gallons, $d_o G^{-x_o}$ calculates per-gallon resource damage, and d_o and x_o are the factors for different oil types.

$$d_o = \begin{cases} 186.69, & \text{gasoline or jet fuel} \\ 274.15, & \text{diesel or LT crude} \\ 279.3, & \text{medium crude, lube oil or IFO} \\ 256.28, & \text{Hvy crude or No. 6 Fuel} \end{cases} \tag{9.20}$$

$$x_o = \begin{cases} 0.1961, & \text{gasoline or jet fuel} \\ 0.1634, & \text{diesel or LT crude} \\ 0.1525, & \text{medium crude, lube oil or IFO} \\ 0.1366, & \text{Hvy crude or No. 6 Fuel} \end{cases} \tag{9.21}$$

$$S_i = \begin{cases} 2.5, & \text{fish} \\ 3.0, & \text{mammals} \\ 4.0, & \text{other high} \\ 4.5, & \text{birds} \\ 3.5, & \text{coral reef} \\ 1.0, & \text{low} \end{cases} \tag{9.22}$$

Similarly, the estimation formulas for shoreline natural resource damage are summarized as:

$$T = (1-e) \sum_{i=1}^{n} V_i = \sum_{i=1}^{n} S_i C d_o C^{-x_o} \tag{9.23}$$

Where C is actual amount that hit the shoreline, and the natural resource modifier S_i is:

Economic and Financial Impacts of Oil Spills

$$S_i = \begin{cases} 4.0, \text{ birds} \\ 2.7, \text{ mammals} \\ 5.0, \text{ endangered species} \\ 4.0, \text{ highly sensitive} \\ 2.0, \text{ moderate sensitive} \\ 0.3, \text{ low sensitive} \end{cases} \quad (9.24)$$

9.2.1.3 Socioeconomic Loss

9.2.1.3.1 Estimation Based on Recovery Function

The oil pollution will also cause losses to various socioeconomic aspects, e.g. fishery and tourism, by reducing their profits. Similar to a contaminated natural resource, the affected economic sectors also need time to recover from the pollution. Their economic losses (EL) are the sum of reduced incomes during their recovery period:

$$EL = \sum_{i=1}^{n} yr_i \sum_{t=0}^{p_i} \left(1 - f_i(t)\right)\left(\frac{1}{1+d}\right)^t \quad (9.25)$$

Where yr_i is the yearly revenue for economic sector i, $f_i(t)$ represents the relative percent of service provided by the affected sector i at year t following the incident, d denotes a yearly discount rate, and p_i quantifies the required period in years for a full recovery (Liu and Wirtz, 2009).

9.2.1.3.2 Estimation Model Based on Oil Spill Database

Similar to the natural resource damage calculation (Etkin and Welch, 2005), the estimation formula for socioeconomic loss can be categorized into on-water and shoreline types. The on-water socioeconomic loss can be summarized as:

$$T = \sum_{i=1}^{n} V_i = \sum_{i=1}^{n} S_i G d_o G^{-x_o} \quad (9.26)$$

Where T is the total value lost, V_i is the lost value for natural resource i, S_i is the socioeconomic resource i modifier, G is the spill amount in gallons, $d_o G^{-x_o}$ calculates per gallon resource damage, and d_o and x_o are the factors for different oil types.

$$d_o = \begin{cases} 1625.6, \text{ gasoline or jet fuel} \\ 2076, \text{ diesel or LT crude} \\ 4407, \text{ medium crude, lube oil or IFO} \\ 1370.8, \text{ Hvy crude or No. 6 Fuel} \end{cases} \quad (9.27)$$

$$x_o = \begin{cases} 0.2166, \text{ gasoline or jet fuel} \\ 0.2226, \text{ diesel or LT crude} \\ 0.2283, \text{ medium crude, lube oil or IFO} \\ 0.2236, \text{ Hvy crude or No. 6 Fuel} \end{cases} \quad (9.28)$$

$$S_i = \begin{cases} 2.9, \text{fisheries} \\ 2.4, \text{ports} \\ 1.0, \text{boating} \\ 1.4, \text{water intake} \\ 2.9, \text{drink water} \\ 1.4, \text{rec fishing} \\ 1.4, \text{other high use} \\ 0.4, \text{low use} \end{cases} \qquad (9.29)$$

Likewise, the estimation formula for shoreline socioeconomic loss can be summarized as:

$$T = (1-e)\sum_{i=1}^{n} V_i = \sum_{i=1}^{n} S_i Cd_o C^{-x_o} \qquad (9.30)$$

Where C is the actual amount that hit the shoreline, and the socioeconomic resource modifier S_i is:

$$S_i = \begin{cases} 1.4, \text{tourist beach} \\ 2.4, \text{wildlife viewing} \\ 0.1, \text{industial} \\ 0.4, \text{port} \\ 1.0, \text{residential} \\ 2.9, \text{high value} \\ 1.0, \text{medium value} \\ 0.4, \text{low value} \end{cases} \qquad (9.31)$$

9.2.1.4 Other Costs

The final category of oil spill impact includes research and legal costs, the value of the lost cargo and tanker or platform, and damage to health, etc. (Grigalunas et al., 1986; Gill et al., 2016). Research costs are the expenditures used for scientists to investigate the influence of oil spills on the environment and society. There are not too many studies on the research cost. The three oil spill research costs reported (Grigalunas et al., 1986; Liu and Wirtz, 2009) are: (1) the *Amoco Cadiz* spill (1978), 3.8 million US dollars; (2) *Exxon Valdez* (1989), 250–270 million US dollars; (3) the *Prestige* spill (2002), 10 million euros.

Legal cost information is viewed as proprietary and thus usually unavailable. Conceptually, it can be measured by summing the value of the time spent by litigants and other marginal legal costs.

The value of the lost cargo and tanker or platform can be estimated according to the market value. However, damage to health is difficult to estimate. The effect on health may come from the air, the water, and the polluted sea food, etc. The short- or long-term effects are usually not obvious. In addition, research studies (IAI, 1990; Gill et al., 2016) show that psychological impacts cause additional public health costs, etc.

9.2.2 Global Economic Impact

The global economic impacts are almost not possible to quantitatively assess, and the impacts can be only visible for very large oil spill accidents. The financial market may be one of the networks which spreads the impacts globally. For example, when the Deepwater Horizon oil spill happened, the BP share price fluctuated immediately and changed with the corresponding oil spill controls and updates, as indicated in Figure 9.6. The company finance and strategies can also be influenced and changed correspondingly. In addition, even oil prices may be impacted. Therefore, all of the relevant industry networks will be affected further in cases of severe oil spill accidents.

In addition, not only may local research costs be caused as mentioned in Section 9.2.1.4, but research costs can also be induced globally. The research and policy may further form new regulations, technologies, and marine activities, which will impact the oil and gas industry, shipping, and shipbuilding industry globally as well.

9.3 MAJOR OIL SPILLS: COMPENSATION, CLAIM, AND COST ESTIMATION

Some major oil spill accidents have happened worldwide. Figure 9.7 shows a distribution of waters with major oil spills from tankers. Some accidents such as the Deepwater

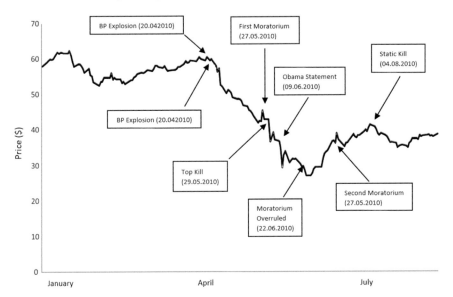

FIGURE 9.6 BP plc share price ($). The figure shows BP PLC daily share prices from January 1, 2010 to September 15, 2010 and significant dates following the BP oil spill explosion (after Sabet et al., 2012).

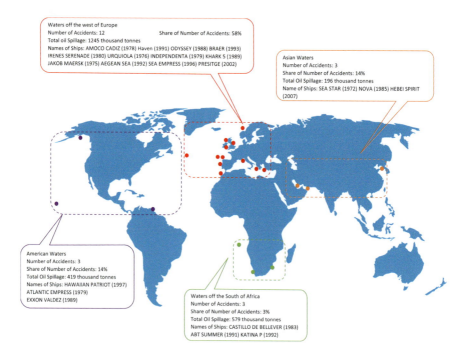

FIGURE 9.7 Distribution of waters with major oil spill accidents from 1970 to 2017 (source: ITOPF, after Chen et al., 2019).

Horizon are thus not listed on the map. This section will firstly illustrate a compensation regime for tankers, i.e. the IOPC Funds briefly and then present some examples of the major accidents in terms of compensations, claims, and estimated cost.

It is known that the original basis of a compensation regime for oil pollution caused by tankers is the 1969 International Convention on Civil Liability for Oil Pollution Damage (CLC, 1969) and the 1971 International Convention on the Establishment of an International Fund for Compensation for Oil Pollution (1971 Fund). Two additional legal instruments, the 1992 Civil Liability Convention (CLC, 1992) and the 1992 Fund Convention (1992 Fund), were created as the amount of compensation required to cover larger oil spill incidents increased. In addition, a third legal instrument, the Supplementary Fund Protocol to the 1992 Fund Convention, was adopted in 2003 to provide additional compensation to affected stakeholders when the compensation payment exceeds the limit under the CLC 1992 and the 1992 Fund. Therefore, the IOPC Funds is now composed of three intergovernmental agreements: the 1971 Fund, the 1992 Fund, and the Supplementary Fund, which provide financial compensation for property damage, oil cleanup operations, and economic losses caused by oil spill incidents occurring in member states (Kim et al., 2014). The 1971 Fund Convention expired in May 2002 but continues to deal with cases of oil spill incidents that occurred before the deadline.

The flag of the vessel which causes the oil spill, the ownership of the oil cargo, and the location of the incident (as long as the damage occurs in a member state)

TABLE 9.10

Damage Estimation and Compensation for Some Oil Spill Incidents (Unit: in USD in Millions)

Case	Size of Oil Spilled (m³)	Damage Estimation Reference	Estimated Damage		Claims	IOPC Funds Limits (CLC + Fund Limit)	Compensation
Amoco Cadiz (1978, France)	~220,000	Thébaud et al., 2005	Global cost of damages	624.4–716.6	681.4		132.5
		Grigalunas et al., 1986	Total net costs to world	195–284			
Tanio (1980, France)	~13,500	Thébaud et al., 2005			160.5		49.7
Exxon Valdez (1989, USA)	45,202	Carson et al., 1992	CVM value	4990.0	2140.0 (including 1250.0 for civil claims + 891 for criminal fines)	N/A (not a member)	6059.0 by Exxon (including 4455.0 for cleanup + 1604.0 for civil settlement by Exxon)
		Carson et al., 2003	CVM value	12,820.0			
		Rodgers Jr et al., 2005	Civil damages, fines, and restitution	2227.0			
		Perry, 2010	Punitive damages (awarded by the jury)	7859.0			
		Allo and Loureiro, 2013	Economic damage	4553.0			
Aegean Sea (1992, Spain)	~80,000	Thébaud et al., 2005			338		16.0
Braer (1993, UK)	~86,500	Thébaud et al., 2005			223.9		83.8
Sea Empress (1996, UK)	83,880	Elliott and LTD, 2001	Direct costs	116.8–138.2	76.1–114.2	122.8	85.9 by IOPC Funds
			Tourism	9.5–10.9			
			Commercial fisheries	16.2–23.8			
			Recreational fisheries	1.9–2.9			
			Human health	2.9–7.1			
			Conservation/non-use	53.5–84.2			
		Thebaud et al., 2005	Global cost of damages	114.3–239.1			
		Allo and Loureiro, 2013	Economic damage	67.5			

(Continued)

TABLE 9.10 (CONTINUED)

Damage Estimation and Compensation for Some Oil Spill Incidents (Unit: in USD in Millions)

Case	Size of Oil Spilled (m³)	Damage Estimation		Claims	IOPC Funds Limits (CLC + Fund Limit)	Compensation
		Reference	Estimated Damage			
Erika (1999, France)	23,067	Thebaud et al., 2005	Global cost of damages	973–1130	264.1	722.5 (including 169.1 by IOPC Funds 553.4 by defendants)
		Bonnieux, 2006	Ecological damage	549.1	507.0	
		IOPC Funds, 2013	Material damage	234.3		
			Moral damage	48.3		
			Pure environmental damage	6.1		
		Allo and Loureiro, 2013	Economic damage	255.7		
Prestige (2002, Spain)	73,628	Loureiro et al., 2006	Commercial and environmental losses	374.0	219.5	1460.1 (including 164.1 by IOPC Funds + 162.7 by European Commission + 1133 by Spanish Government)
			Cleaning up, recovery, and other palliative measurements	729.5	1494.0	
			Other expenditures including transfer to fishermen and shellfish pickers	326.9		
		Garza-Gil et al., 2006	Losses in coastal fishing sectors	37.4		
			Losses in aquaculture	7.9		
		Garza et al., 2009	Cleanup (marketed)	800.6		
			CVM value	1763–2240		
			Fisheries (marketed)	976.8		
			Tourism (marketed)	144.6		
			Recreation (non-marketed)	124.5		
			Biodiversity (non-marketed)	1236.3		
		Loureiro et al., 2009	CVM value	833.9		
		Allo and Loureiro, 2013		1481.8		

(Continued)

TABLE 9.10 (CONTINUED)

Damage Estimation and Compensation for Some Oil Spill Incidents (Unit: in USD in Millions)

Case	Size of Oil Spilled (m³)	Damage Estimation		Claims	IOPC Funds Limits (CLC + Fund Limit)	Compensation
		Reference	Estimated Damage			
Hebei Spirit (2007, Korea)	12,547	N/A		3990.0	300.0	1600.0
Deepwater Horizon (2010, Gulf of Mexico)	~780,000	Smith et al., 2010	Estimated costs 36,900			
		Lee et al., 2018; BP annual report, 2010–2016	Net costs 62,585			
		Lee et al., 2018	Ultimate costs 145,925			

Source: after Kim et al. (2014) and Thébaud et al. (2005). In Thébaud et al. (2004) 1 £ = 1.45 $ is used to transfer the reported numbers in year 2001.

are not related to the provision of financial compensation for oil pollution under the CLC and the 1992 Fund (IOPC Funds, 2014). First, the costs are paid by the individual tanker owner up to the relevant CLC limit through compulsory insurance and then by IOPC Funds. If the accepted claims exceed the Fund limit, every claimant is treated equally and receives a corresponding proportion of their claims, and the Supplementary Fund will work for its relevant members to provide additional compensation to affected stakeholders.

Table 9.10 summarizes compensations, claims, and estimated costs from different sources for some major oil spill accidents in Figure 9.7. It can be seen that a relatively low rate of compensation is paid by IOPC Funds, which may be because non-use values were disregarded in the damage assessment. Meanwhile, there are also difficulties in giving proofs to establish the damage incurred by oil spill pollution (Kim et al., 2014). The compensation rate is high in the case of the *Exxon Valdez* oil spill. This is because the United States legal system did not permit limitation of liability, contrary to the IOPC Funds regime, which allows compensation for the non-use values of the natural resource (Carson et al., 2003; Kim et al., 2014). In the USA, the oil pollution damage has been regulated by Oil Pollution Act of 1990, enacted after the *Exxon Valdez* oil spill.

In general, from Table 9.10, it can be seen that major oil spills usually cause huge economic losses and the claims cannot be met by IOPC Funds solely. Some governmental involvement makes the compensations higher. In some cases, the estimated costs may somehow reflect the economic damages, but this varies a lot for different situations. More efforts and studies are still needed to understand the impacts better.

9.4 SUMMARY

This chapter discussed the economic and financial impacts of oil spills. First, oil, oil spills, and response and recovery systems are described and offshore oil spills are identified as main targets in this chapter. Second, subdivisions of oil spill economic and financial impacts are introduced and estimation methods and models are summarized. Last, some major oil spill accidents are summarized with the compensation, claims, and estimated costs.

The major oil spills usually cause large impacts on the ecology and society, and the compensations are often not sufficient. Additionally, the impacts may last for a long period. Therefore oil spill prevention and preparedness are always important to avoid and mitigate the oil spill and its economic and environmental impacts. Meanwhile, new challenges from more marine activities in ice-covered regions may require more emphasis in both technical and economic research.

REFERENCES

Allo, M. and M.L. Loureiro. 2013. Estimating a meta-damage regression model for large accidental oil spills. *Ecological Economics,* 86, pp. 167–175.

Bonnieux, F. 2006. *Evaluation économique du préjudice écologique causé par le naufrage de l'Erika.* Rapport Confidentiel.

Economic and Financial Impacts of Oil Spills

Carson, R.T., R.C. Mitchell, M. Hanemann, R.J. Kopp, S. Presser and P.A. Ruud. 1992. *A contingent valuation study of lost passive use values resulting from the Exxon-Valdez oil spills.* Report to the Attorney General of Alaska.

Carson, R.T., R.C. Mitchell, M. Hanemann, R.J. Kopp, S. Presser, P.A. Ruud. 2003. Contingent valuation and lost passive use: Damages from the Exxon Valdez oil spill. *Environmental and Resource Economics*, 25, pp. 257–286.

Chang, S.E., J. Stone, K. Demes and M. Piscitelli. 2014. Consequences of oil spills: A review and framework for informing planning. *Ecology and Society*, 19(2), p. 26.

Chen, J., W. Zhang, Z. Wan, S. Li, T. Huang and Y. Fei. 2019. Oil spills from global tankers: Status review and future governance. *Journal of Cleaner Production*, 227, pp. 20–32.

Devold, H. 2013. *Oil and gas production handbook: An introduction to oil and gas production, transport, refining and petrochemical industry.*ABB Oil and Gas, https://library.e.abb.com/public/34d5b70e18f7d6c8c1257be500438ac3/Oil%20and%20gas%20production%20handbook%20ed3x0_web.pdf.

Elliott, M. and P. LTD. 2001. *Economic for the Environment Consultancy LTD (EFTEC): Study on the valuation and restoration of damage to natural resources for the purpose of environmental liability.* European Commission, Directorate-General Environment, Brüssel.

EPPR. 2015. *Guide to oil spill response in snow and ice conditions.* Emergency Prevention, Preparedness and Response (EPPR),Arctic Council.

Etkin, D.S. 1999. Estimating cleanup costs for oil spills. *Proceedings of 1999 international oil spill conference*, pp. 35–39.

Etkin, D.S. 2000. Worldwide analysis of oil spill cleanup cost factors. *Proceedings of 23rd Arctic & Marine oilspill program technical seminar*, pp. 161–174.

Etkin, D.S. and J. Welch. 2005. *Oil spill response cost-effectiveness analytical tool (OSRCEAT).* Environmental Research Consulting CICEET.

Garza-Gil, M.D., J.C. Surís-Regueiro and M.M. Varela-Lafuente. 2006. Assessment of economic damages from the Prestige oil spill. *Marine Policy*, 30(5), pp. 544–551.

Garza, M.D., A. Prada, M. Varela and M.X.V. Rodríguez. 2009. Indirect assessment of economic damages from the Prestige oil spill: Consequences for liability and risk prevention. *Disasters*, 33, pp. 95–109.

Gill, D.A., L.A. Ritchie and J.S. Picou. 2016. Sociocultural and psychosocial impacts of the exxon valdez oil spill: Twenty-four years of research in Cordova, Alaska. *The Extractive Industries and Society*, 3, pp. 1105–1116.

Grigalunas, T.A., R.C. Anderson, G.M. Brown, R. Congar, N.F. Meade and P.E. Sorensen. 1986. Estimating the cost of oil spills: Lessons from the Amoco Cadiz incident. *Marine Resource Economics*, 2, 239–262.

Helle, I., H. Ahtiainen, E. Luoma, M. Hänninen and S. Kuikka. 2015. A probabilistic approach for cost-benefit analysis of oil spill management under uncertainty: A Bayesian network model for the Gulf of Finland. *Journal of Environmental Management*, 158, pp. 122–132.

Impact Assessment, Inc. (IAI), 1990. Economic, social and psychological impact assessment of the Exxon Valdez Oil Spill. Final Report Prepared for Oiled Mayors Subcommittee, Alaska Conference of Mayors, Anchorage, Alaska.

Manual on oil spill risk evaluation and assessment of response preparedness. International Maritime Organization.

IOPC Funds. 2013. *Incidents involving the IOPC funds 2012.* International Oil Pollution Compensation Funds, London, United Kingdom.

IOPC Funds. 2014. *The international regime for compensation for oil pollution damage: Explanatory note.* International Oil Pollution Compensation Funds, London, pp. 1–11.

Kim, D., G. Yang, S. Min and C. Koh. 2014. Social and ecological impacts of the Hebei Spirit oil spill on the west coast of Korea: Implications for compensation and recovery. *Ocean & Coastal Management*, 102, pp. 533–544.

Kontovas, C.A., H.N. Psataftis and N.P. Ventikos. 2010. An empirical analysis of IOPCF oil cost data. *Marine Pollution Bulletin*, 60, pp. 1455–1466.

Lee, Y.G., X. Garza-Gomez and R.M. Lee. 2018. Ultimate costs of the disaster: Seven years after the deepwater horizon oil spill. *The Journal of Corporate Accounting & Finance*, 29(1), pp. 69–79.

Liu, X. and K.W. Wirtz. 2006. Total oil spill costs and compensations. *Maritime Policy & Management*, 33, pp. 49–60.

Liu, X. and K.W. Wirtz. 2009. The economy of oil spills: Direct and indirect costs as a function of spill size. *Journal of Hazardous Materials*, 171, pp. 471–477.

Loureiro, M.L., J.B. Loomis and M.X. Vazquez. 2009. Economic valuation of environmental damages due to the prestige oil spill in Spain. *Environmental and Resource Economics*, 44(4), pp. 537–553.

Loureiro, M.L., A. Ribas, E. Lopez and E. Ojea. 2006. Estimated costs and admissible claims linked to the Prestige oil spill. *Ecological Economics,* 59(1), pp. 48–63.

Lu, L., F. Goerlandt, O.A. Valdez Banda, P. Kujala, A. Höglund and L. Arneborg. 2019. A Bayesian Network risk model for assessing oil spill recovery effectiveness in the ice-covered Northern Baltic Sea. *Marine Pollution Bulletin*, 139, pp. 440–458.

Lu, L., Goerlandt, F., Tabri, K., Höglund, A., Valdez Banda, O. A., & Kujala, P. (2020). Critical aspects for collision induced oil spill response and recovery system in ice conditions: A model-based analysis. *Journal of Loss Prevention in the Process Industries*, 66, 104198.

Montewka, J., M. Weckström and P. Kujala. 2013. A probabilistic model estimating oil spill clean-up costs – A case study for the Gulf of Finland. *Marine Pollution Bulletin*, 76, pp. 61–71.

NOAA. 1999. *Discounting and the treatment of uncertainty in natural resource damage assessment.* National Oceanic and Atmospheric Administration Damage Assessment and Restoration Program, Washington, DC.

Oil & Gas UK, OPOL. 2012. *Oil spill cost study-OPOL financial limits.* Oil & Gas UK.

Perry, R. 2010. Economic loss, punitive damages, and the Exxon Valdez litigation. *Georgia Law Review*, 45, p. 409.

Rodgers Jr., W.H., J. Crosetto III, C. Holley and T. Kade. 2005. Exxon Valdez reopener: Natural resources damage settlements and roads not taken. *Alaska Law Review*, 22, p. 135.

Sabet, S.A.H., M.A. Cam and R. Heaney. 2012. Share market reaction to the BP oil spill and the US government moratorium on exploration. *Australian Journal of Management*, 37(1), pp. 61–76.

Shahriari, M. and A. Frost. 2008. Oil spill cleanup cost estimation-Developing a mathematical model for marine environment. *Process Safety and Environmental Protection*, 86(3), pp. 189–197.

Smith, L. C., Smith, M., & Ashcroft, P. (2010). Analysis of Environmental and Economic Damages from British Petroleum's Deepwater Horizon Oil Spill. *Albany Law Review*, 74 (1), 563–586.

Thébaud, O., D. Bailly, J. Hay and J.P. Agundez. 2005. *The cost of oil pollution at sea: An analysis of the process of damage valuation and compensation following oil spills.* Economic, Social and Environmental Effects of the Prestige Oil Spill de Compostella, Santiago, pp. 187–219.

Turner, R.K., S.E. Subak and W.N. Adger. 1995, *Pressure, trends and impacts in the coastal zone: Interations between socio-economic and natural system.* CSERDE Working Paper.

Index

A

ABC News, 10
Absorbent materials, 8
Absorbents, 8
ABT Summer, 85, 87
Acentric factor, 24, 29
Advanced simulators, 12
Advection, 12
Aegean Captain, 5
Aggregation, 202, 203
Air and wind speed, 207
Aircrafts, 8
Alabama, 6
Alaska oil spill, 203
Allision, 64
Amoco Cadiz, 5, 85, 88
ANSYS, 12
API gravity, 217
Application of the model to the case of
continuous flow of oil, 217–220
Approximate simulations, 12
Aquatic environments, 203
Area of oil slick, 220
Aromatic compounds, 21
Aromatic hydrocarbons, 203
Aromatics, 20
Asphaltene, 21, 202
Atlantic Empress, 5
Automatic Oil Spill Detection, 200
Autonomous underwater vehicles (AUVs), 269

B

Barbados, 5
Bay Jimmy, Louisiana, 7
Bay of Campeche, Gulf of Mexico, 5
Behavior of an oil spill on seawater surface, 9
Biden, Joe, 157, 158, 160
Biodegradation, 170–171, 201, 255
Bioemulsifiers, 8
Biofuels, 52, 204
worldwide biofuels production, 55
Bioremediation, 8, 10
Biosurfactants, 8
Birds and fish, oil spill impact, 136–139
BLOSOM, 13
Blowout preventer (BOP) failure, 102
Boats, 8
Boiling point, 37, 207
Booms, 8

BP oil spill, 109, 113, 115, 164
birds and fish, impact on
birds struggling with oil, at Louisiana
coast, 138
dolphin in oilwater, in Chandeleur Sound,
La., 143
oiled baby brown pelican chick, 137
oiled pelicans, in Gulf of Mexico, 136
seabird, soaked in oil, 139
volunteer rescues brown pelican, at Fort
Jackson Rehabilitation Center, 137
BP-GOM, 10
distribution of, 196
economic impacts
lower marine riser package (LMRP), 154
operator and principal, 149
top kill procedure, 153
history of, 155–157
oil spill trajectory, 11
political impacts, 157–161
slick thickness, 204
shores and people, impact of
benzene and aromatics, in air, 146
2-butoxyethanol, 149
chemical dispersants, 149
dolphin death, 145
HSE worker, 148
Louisiana wetland inhabitants, 141
in Louisiana's marshlands, 143
in marsh grass, Barataria Bay, 142
oil-covered crab, on beach of Grand Terre
Island, 141
oil-dispersing chemical, 148
in Orange Beach, 146
seabird mortalities, 144
simulated extent, of oil slick, 140
toxic materials, 149
US Coast Guard vessel, 149
statistical report, 5British Petroleum (BP), 5, 6,
9, 201, 222
Brown Pelican, 6
Bubbles, 13
Buras, Louisiana, 137
Burning, 201
2-Butoxyethanol, 149

C

Calculation procedure, 207–211
Cape Town, South Africa, 5

293

294

Index

Castillo de Bellver, 5, 85, 87
Centers for Disease Control and Prevention
  (CDC), 146
CEO, 7
Chandeleur Islands, 137
Chandeleur Sound, 140
Chapman–Enskog equation, 40
Chemical agents, 201
Chemical dispersants, 149, 256
Chemical dispersion, 47
Chemical methods, 8, 189
Clean Kool, 164
Clean up operation, 201, 204
Cleanup costs, 14
  Cleanup Cost Study in Finland, 273–275
  estimation model, 271
  Etkin Model
    general cleanup cost, 275, 277, 278
    on-water cleanup cost, 278
    shoreline cleanup cost, 278–280
  limit-parameter-based regression model, 280
  Oil Spill Cost Study in UK, 271–273
  regression model, 280
Cleanup methods
  deployment, 165
  dispersants
    feasibility of, 188
    guidelines for application, 187
    high oil viscosity, 187
    oil concentration, in water, 188
    oil viscosity, 187
    surfactants, 182
    treatment *vs.* types of accident and spilled
      oil, 189
    type of dispersant, 183
    type of spraying method and treatment
      rate, 186
    type of surfactants and solvents, 185
  in-situ burning
    airplane releasing oil dispersan, 182
    Deepwater Horizon/BP oil spill, 180, 181
    drillship Discoverer Enterprise, 182
    DWH/BP/GOM oil spill, 180
    fire-resistant booms, 178
    ignition devices, 179
    smoke billow up, 179
  natural processes
    biodegradation, 170–171
    dispersion, 167, 169
    emulsification, 169, 170
    evaporation, 166–167
    oxidation, 169, 170
    sedimentation, 169–171
    spreading, 167–168
  protective booms
    decontamination site, in Venice, 176

deflection, 172
deployment oF, 173
ITOPF, 172
Naval Air Station Pensacola Pollution
  Response unit, 178
oil containment and concentration, 172
oil containment boom at Pensacola, 177
oil-absorbent boom in warehouse, 174
protection, 172
re-deployment, 174
shrimp boats tow fire-resistant oil-
  containment boom, 176
US Coast Guard, 178
workers pressure, 174
skimming and mechanical removal methods
  "A Whale" skimmer ship, 192
  beach cleanup workers, 193
  cleanup crew member, in Grand Isle,
    La., 193
  DWH oil spill on Louisiana beaches, 190
  heavy machinery, to remove oil in Puerto
    Rico, 195
  oil mechanical and chemical oil spill
    response, 190
  oil-impacted beaches north of Santa
    Barbara, 194
  sand berms, 192
  self-propelled weir skimmer, 192
  time trajectories of oil volumes, 196
Cleanup operation, 14, 147
CNN, 5, 164
Coagulation, 8
Coal liquids, 204
Coats marsh, 7
Collision, 64
Commercial models, 14
Comparison of model predictions with lab
  experiments, 212
Compensation
Comprehensive models, 12
Compressibility factor, 206
Computational fluid dynamics (CFD), 12
Concentration of hydrocarbon type, 12
Continuous flow of oil into sea, 217–220
Continuous oil flow
  GC analysis, 217
  GOM oil spill, 221
  hydrocarbons, molar distribution of, 218
  low and high temperature, 219
  oil API gravity, 217
  oil vaporized *vs.* time, 221
  temperature *vs.* time, 219
  wind speed vs time, 220
Corrosion, 4
Critical pressure, 24, 26
Critical temperature, 24, 42

Index

295

Crude oil, 1, 48, 201, 204, 208, 210, 211, 214, 216, 222
  API gravity and viscosity, 22
  aromatic compounds, 21
  asphaltene, 21
  asphaltene molecule, 21
  characterization of, 3
  classification of, 22
  flashed gas, 19
  gas-to-oil ratio, 19
  hydrocarbons, 20
  iso-octane (2-methylheptane), 20
  methyl groups, 21
  mono-olefins, 20
  naphthenes and aromatics, 20
  naphthenic group, 20
  n-Butane, 20
  offshore production and transportation, 48
  oil refinery, 22
  olefin-free, 20
  paraffins, 20
  petroleum fractions, 22
  polynuclear aromatic hydrocarbons (PAHs), 21
  properties of
    in ASTM Manual 50, 29
    density and specific gravity, 30
    kinematic viscosity, 31
    Lee–Kesler method, 29
    refractive index, 30
    single carbon number (SCN) group, 31
  and reservoir fluids, 19, 32
  sulfur and nitrogen, 21
Currents, 13

D

Damage estimation, 14
Dead weight tons (dwt), 1, 60
Decomposition of oil, 201
Deepwater Horizon, 1, 85, 86, 114, 144, 201
  aerial image of, 109
  blowout preventer (BOP) failure, 102
  chronology of events, 99–101, 104
  location of explosion, 104
  Macondo well, 102
  Mississippi Delta, 102, 108
  rig's blowout preventer, 105
Degradation of oil, 201
Degradation processes, 10
Degree of toxicity of dissolved components of oil
    in water, 211
Demulsifier, 8
Density, 9, 35
  of gases, 34
  for pure liquid, 35
Deployment, 165

Detection of oil pollution, 200
DHI MIKE user interface, 248
Diffusion coefficient, 3, 206
Dimensionless properties, 206
Dispersants, 5, 8
Dispersants, cleanup methods
  feasibility of, 188
  guidelines for application, 187
  high oil viscosity, 187
  oil concentration, in water, 188
  oil viscosity, 187
  solvents, 185
  spraying method and treatment rate, 186
  surfactants, 182, 185
  treatment *vs.* types of accident and spilled
      oil, 189
  type of, 183
Dispersion, 167, 169, 201, 222
Dissolution, 10, 12, 201, 254–255
Dissolved compounds, 13
Distribution of heavy components, 12
Droplets, 13
Dynamic processes, 203

E

Economic, 6
Economic impacts, oil spill
  lower marine riser package (LMRP), 154
  operator and principal, 149
  top kill procedure, 153
Emulsification, 10, 169, 170, 201, 253–254
Entrained seawater, 13
Environmental impacts, 1
EPA, 9
Estimating rate of oil vaporization, 218
Ethane, 204
Etkin Model
  general cleanup cost, 275, 277, 278
  on-water cleanup cost, 278
  shoreline cleanup cost, 278–280
Evaporation, 10, 12, 166–167, 201, 202, 204, 207, 217, 252–253
Excel program, 218
Experimental data, 204
Exploratory well blowout, 5
Exxon Valdez oil spill, 10, 64, 90, 145

F

Fate of oil spill, 10, 200
Federal agency, 201
Fergana Valley, 5, 85–87
Filtration, 8
Final fate of BP oil spill in Gulf of Mexico by
    NOAA model, 222

296 Index

Financial markets, 14
Finland, 14
Flash Point, 3
Floating production storage and offloading
    (FPSO) system, 57
Flocculation, 8
Florida, 6
Flotation, 8
Flux of dissolution, 206
Fuels, 204

G

Gas constant, 206
Gas hydrates, 13
Gas, oil, and reservoir fluid, 117
Gas to oil ratio (GOR), 19, 117, 217
Gasoline, 48, 201
General NOAA Operational Modeling
    Environment (GNOME), 13
Genoa, Italy, 1
Geographic information system (GIS), 84
Geographic positioning system (GPS), 84
GIS (geographic information system), 200
Glaso's correlation, 39
GOM oil spill, 6, 201, 221
Graphical interface, 14
Gravity separation, 8
Gulf of Mexico (GOM), 9, 201
Gulf war oil spill, 200

H

Haven, 1, 85, 88
Heavy compounds, 201
Henry's constant, 44
Henry's constant for light gases, 44
Henry's law, 43
Heptane plus fraction, 208
High oil viscosity, 187
Horizontal turbulent diffusion, 12
Human error, 1
Hydrocarbons, 20
    properties of, 27–28
        acentric factor, 24
        API gravity, 23
        boiling point, 23
        carbon-to-hydrogen weight ratio, 25
        critical compressibility factor, 24
        critical pressure, 24
        critical temperature, 24
        critical volume, 24
        density, 23
        flash point, 25
        freezing point, 25
        molecular weight, 23

        refractive index, 26
        specific gravity, 23
        specific volume, 23
        UOP characterization factor, 25
        vapor pressure, 24
    solubility of
        Henry's constant, 44
        liquid–liquid equilibria, principle of, 42
Hydrodynamic models, 245

I

Ideal oil spill model, 203
IEA, 1
Industry network, 14
Initial oil spill surface area, 211
Initial oil spill volume, 212
Initial volume of spill, 10
Inorganic sorbents, 8
In-situ burning, 8, 164
    airplane releasing oil dispersan, 182
    Deepwater Horizon/BP oil spill, 180, 181
    drillship Discoverer Enterprise, 182
    DWH/BP/GOM oil spill, 180
    fire-resistant booms, 178
    ignition devices, 179
    smoke billow up, 179
International Bird Rescue Research Center,
    136, 137
Iraq-Iran war, 87
Iso-octane (2-methylheptane), 20
Italy Milan, 3
ITOPF, 8
Ixtoc I, 85, 86
Ixtoc I oil spill, 5

K

Kerosene, 216
Kinematic viscosity, 31
Kouzel correlation, 40
Kuwaiti crude oil for export, 210
Kuwait–Iraq war, 1

L

Laboratory experiments, 211–217
Large crude carriers (LCC), 1
Largest oil spills, 3
Lee–Kesler method, 29
Light oils, 200
Light petroleum fractions, 201
Liquefied natural gas (LNG), 1, 52, 204
    imports, 54
    worldwide major trade movements for, 55
liquefied natural gas or LNG, 204

Index

Liquid chemicals, 204
Liquid compressibility factor, 206
Liquid density, 206
Liquid mixtures, 36
Liquid–liquid equilibria (LLE), 42
Louisiana, 6
Louisiana wetland inhabitants, 141
Lower marine riser package (LMRP), 154
LPG, 1
Lumped hexanes, 208

M

Macondo well, 6, 102
Maintenance, 4
Man-made cleaning technique, 163
Marine environment, 201
Marine Tankers, 5
Marine transportation, in oil spill occurrence
 by natural seeps, 69–71
 by pipelines, 65–68
 by refineries, 69
 by tanker
  allision, 64
  collision, 64
  dead weight (DWT), 60
  Exxon Valdez oil spill, 64
  the Sanchi, 65
  in Strait of Hormuz, 62, 63
  tanker sizes and uses, 61
  very large crude carriers (VLCC), 60
  world oil production, 60
 by war, sabotage, and actions, 71–76
Marine-type oil spills, 10
Maseela Beach, 211
Mass of oil, 204, 207
 dissolved in water, 207
Mass transfer coefficient, 207
Mass transfer coefficient
 in air for vaporization, 207
Mathematical modeling, 203
Mechanical recovery, 201, 256, 269
Methyl groups, 21
Microbes, 8
MIKE, 13
Mississippi Delta, 105, 108
Mississippi River, 6, 117
Model minimum input data, 241
Model prediction
 for the amount of oil vaporized *versus* time
  for BP oil spill, 221
 for the mass of crude oil and aromatic
  hydrocarbons dissolved in water, 215
 for the mass of crude oil spill remaining on
  the water, 214
 for the rate of oil crude dissolution, 216
 for the rate of oil dissolution, 215

Modeling and simulation methods, 12
Modeling scheme, 204–207
Models, 2, 10, 12, 13
MOHID, 13
Molar distribution of hydrocarbons in the oil, 218
Molecular weight, 201
 of oil, 215
Mono aromatic hydrocarbons, 12
Mono-aromatics, 12
Mono-olefins, 20
Multi-phase volume, 13

N

Naphthenes, 20
Naphthenic group, 20
Narrow-cut hydrocarbon mixtures, 208
National Oceanic and Atmospheric
  Administration (NOAA), 6
Natural disaster, 4
Natural dispersion, 10, 12
Natural gas, 1, 52, 53
 percentage distribution of, 53
 production and consumption by region, 54
 reserves-to-production (R/P), 53
Natural processes, cleanup methods
 biodegradation, 170–171
 dispersion, 167, 169
 emulsification, 169, 170
 evaporation, 166–167
 oxidation, 169, 170
 sedimentation, 169–171
 spreading, 167–168
natural resource damages, 14
Navier–Stokes, 12
Navier–Stokes equation, 244
n-Butane, 20
Newfoundland, Canada, 5
NOAA, 6
Nowruz oil field, 5, 85, 87, 95
Number of moles, 205

O

Obama, Barack, 157–160
Odyssey tanker, 5, 85, 88
Offshore Activity, 4
Offshore oil production, 56
 and gas production, 56
 share of, 56
Offshore oil spills
 financial aspects of, 14
Oil behavior at seawater surface, 202
Oil composition, 203
Oil dissolution, 201
Oil floating on the sea, 125
Oil jets, 13

Oil locations, 112
Oil molar concentration, 205
Oil molar density, 207
Oil plume dynamics and underwater degradation processes
  Deepwater Horizon blowout, 236
  high turbulent kinetic energy and momentum, 235
  hydrocarbon mixture, typical phase envelope of, 234
  methane hydrate phase diagram, 235
  oil droplets and bubbles, 236
  OpenDrift-TAMOC model's user interface, 239, 241
  phase transition and buoyancy, 236
  plume simulation, 237
  TAMOC near-field model, 238
Oil pollution, 14
  socioeconomic aspects, 14
Oil production and consumption, 51
Oil refinery, 22
Oil removal with the use of absorbents, 201
Oil reserves
  percentage distribution of, 50
Oil sedimentation, estimating rate of, 214
Oil solubility in water, 201, 208
Oil spill
  arial view, 4
  ABT Summer, 85, 87
  air humidity *vs.* days, 121
  Amoco Cadiz, 85, 88
  Atlantic Empress and *Aegean Captain*, 85, 88
  behavior of, 10
  BP-GOM oil spill, 4, 6, 10, 11
  BP oil spill, 109, 113, 115
  Castillo de Bellver, 85, 87
  chemical treatment of, 10
  and cleanup methods. (*see* Cleanup methods)
  compensation, claim, and cost estimation, 285–290
  Deepwater Horizon, 85, 86, 114
    aerial image of, 109
    blowout preventer (BOP) failure, 102
    chronology of events, 99–101, 104
    location of explosion, 104
    Macondo well, 102
    Mississippi Delta, 102, 108
    rig's blowout preventer, 105
  defined, 1
  detection, 84–85
  in Europe, 83
  Exxon Valdez oil spill, 64, 90
  failed blowout preventer, 106
  Fergana Valley, 85–87
  forecasting, 225
  gas, oil, and reservoir fluid, 117

gas-to-oil ratio (GOR), 117
global economic impact, 285
in Gulf of Mexico, 83, 109, 111, 116
Haven, 85, 88
in history, 1
in-situ burning of, 256
Ixtoc I, 85, 86
liquefied natural gas (LNG), 1
local economic impact
  cleanup costs, 271–280
  natural resource damages, 280–283
  research and legal costs, 284
  socioeconomic loss, 283–284
in Louisiana, 7
marine transportation
  by natural seeps, 69–71
  by pipelines, 65–68
  by refineries, 69
  by tanker, 59–65
  by war, sabotage, and other actions, 71–76
marine-type oil spills, 10
in Milan, 3
Mississippi Delta, 108
Nowruz oil field, 85, 87
occurrence and causes of, 5
Odyssey tanker, 85, 88
oil and gas system, 266–267
oil locations, 112
oil spill response and recovery system, 268–269
by oil tankers, 82
Persian Gulf, 5
petroleum exploration and production
  floating production storage and offloading (FPSO) system, 57, 58
  gas production, 56
  offshore oil production, 56
  oil and hydrocarbons, 59
Production Well, D-103, 85, 89
potential oil spills, 267–268
reservoir formation pressure, 117
Sanchi tanker, 85, 89
SAR image classification, 85
on seawater surface, 9
simulation and prediction of fate, 201
since 1989, 83
size and cause of, 5
size of tankers, 1
skimming, 10
source of, 3
spreading, horizontal movement, 12
SS Atlantic Empress collision, 90
surface area, 122–125
temperature, humidity, wind speed, and cloud overage, 118–121
3D oil spill modeling, 12

Index

trajectory and simulation, 126–130
UN Glossary of Statistical Terms, defined, 47
US oil spill, 106
in West Africa, 82
Oil spill advection
analysis and reanalysis data, 248
DHI MIKE user interface, 248
free, open-source, and commercial far-field oil spill models, 246
hydrodynamic models, 245
Navier–Stokes equation, 244
OpenDrift-TAMOC user interface, 249
open-source and commercial models, 244
water current and wind inputs, 247
Oil spill disappearance
aggregation, 202
aromatic hydrocarbons, 203
calculation procedure, 207–211
continuous oil flow
GC analysis, 217
GOM oil spill, 221
hydrocarbons, molar distribution of, 218
low and high temperature, 219
oil API gravity, 217
oil vaporized vs. time, 221
temperature vs. time, 219
wind speed vs time, 220
degradation, 201
field experiments, 203
Gulf of Mexico (GOM) oil spill, 201
heavy compounds and residues, 203
laboratory experiments and model predictions, 211–217
liquefied natural gas (LNG), 204
modeling scheme, 204–207
position and area, 220
SAR image classification, 200
on seawater surface, 202
temperature-dependent mass transfer coefficients, 204
weather conditions, 200
Oil spill modeling and simulation techniques
far-field oil spill modeling
critical inputs and processes, 249–250
oil entrainment and droplet size distribution, 250–251
oil spill advection, 244–249
statistical analysis for contingency planning, 256–260
wave data, 249
weathering and oil decaying processes, 251–256
near-field and far-field oil spill modeling, 226–227
oil spill forecasting, 225
scientific studies, 225

statistical studies, 225
subsea blowouts and near-field modeling
equilibrium droplet size model, 231–232
oil plume dynamics and underwater degradation processes, 234–241
physical processes, 228–230
population dynamic droplet size model, 233
reservoir fluid vs. surface oil, 227
typical input data, 241–242
Oil spill occurrence, 1
Oil spill surface area, 252
Oil spill thickness, 250, 251
Oil spill volume, 8
Oil tanker
attacked, 2
UK oil tanker, 2
Oil trade movements, 50, 52
Oil viscosity, 41, 183, 187
OILFOW2D, 12
OILMAP, 13
Olefin-free, 20
Oleophilic material, 8
OpenDrift-TAMOC model's user interface, 239, 241
OpenDrift-TAMOC post-processing user interface, 259
OpenDrift-TAMOC user interface, 249
Open-source models, 14
Organic sorbents, 8
OSCAR, 12
OSCAR DeepBlow, 13
Oxidation, 170
oil, 169

P

Paraffins, 20
Persian Gulf oil spill, 1, 5, 95–97
Petrochemicals, 204, 267
Petroleum exploration and production
floating production storage and offloading (FPSO) system, 57, 58
offshore oil and gas production, 56
oil and hydrocarbons, 59
share of offshore oil production, 56
Petroleum fraction, 3
Petroleum industry, 48
Petroleum mixtures, 32–33
Petroleum non-fuels, 204
Petroleum products, 3, 204
characterization of, 3
Photooxidation, 256
Physical methods, 8
Physiochemical conditions, 201
Pipelines, 65–68

300   Index

Plume, 13
Plume simulation, 13
Poly aromatic hydrocarbons, 211
Polynuclear aromatic hydrocarbons (PAHs), 21
Portsall, France, 5
Position of BP oil spill *versus* time, 220
Pour Point, 33–37
Prediction of oil slick area, thickness and volume
    versus time, 220
Prediction of oil toxicity in water, 201
Probability density function for a property, 32
Production Well, D-103, 5, 85, 89
Propane, 204
Pseudocomponent properties, 205–207
Pumps, 8
Pure hydrocarbons, 23–26

R

Rackett equation, 36, 206
Randolf, Charlotte, 158
Rate of change of slick thickness for Kuwaiti
    crude, 213
Rate of dissolution, 215
    of monoaromatics in water for naphtha
        product, 215
    of monoaromatics in water from a kerosine
        oil spill, 217
Rate of mass transfer flux, 205
Rate of oil disappearance, 203
Rate of vaporization, 24
Re-deployment, 174
Reduced parameter, 206
Refineries, 69
Reflection of sun, 6
Refractive index, 30, 210
Remote operation vehicles (ROVs), 269
Reserves-to-production (R/P) ratios
    natural gas, 53
    by region, 51
Reservoir fluids, 32
Residues, 203
Response and Restoration's Emergency Response
    Division, 13
Response vessels, 8
Reuters, 65
Russia Today, 1

S

Salt concentration in sea water, 208
Sanchi oil spill, 90–91
SAR (Synthetic Aperture Radar), 200
Saturated liquids, 36
Sea of Oman, 1
Seabird mortalities, 144

Seawater surface, 202
Sedimentation, 10, 12, 169–171, 201, 255–256
    oil components, 207
Selecting cleanup method, 201
Semi-analytical approach, 204
Semi-analytical models, 220
Shipboard oil pollution emergency plan
    (SOPEP), 268
Shoreline costs, 14
Shores and people, oil spill impact
    benzene and aromatics, in air, 146
    2-butoxyethanol, 149
    chemical dispersants, 149
    dolphin death, 145
    HSE worker, 148
    Louisiana wetland inhabitants, 141
    in Louisiana's marshlands, 143
    in marsh grass, Barataria Bay, 142
    oil-covered crab, on beach of Grand Terre
        Island, 141
    oil-dispersing chemical, 148
    in Orange Beach, 146
    seabird mortalities, 144
    simulated extent, of oil slick, 140
    toxic materials, 149
    US Coast Guard vessel, 149
Simple models, 199–222
Simulation animation, 13
Single carbon number (SCN) group, 31
Size of oil spill, 200
Skimmers, 8, 142, 201
Skimming, 10
Slick thickness, 10
Smoke rises from a controlled burn, 1
Socioeconomic, 14
Solubility of component of oil in water, 203
Solubility of heavy and light components, 201
Solubility of mono-aromatics (MA), 211
Solubility of oil in water at presence of
    salt, 215
Solubility of poly-aromatics (PA), 211
Solubility Parameter, 3
Specific gravity, 212
Spill response equipment, 8
Spill trajectory, 203
Spray systems, 8
Spreading, 9–10, 167–168, 203, 251
    of Exxon oil spill, 203
    of oil, 201
SS Atlantic Empress collision, 90
Static kill, 6
Storage, 8
Storm breaking tanker, 5
Strait of Hormuz, 1
Subsea, 13
Subsea blowouts and near-field modeling

Index

301

equilibrium droplet size model, 231–232
equilibrium models, 230
oil plume dynamics and underwater
    degradation processes, 234–241
physical processes, 228–230
population dynamic droplet size model, 233
reservoir fluid *vs.* surface oil, 227
typical input data, 241–242
Surface oil spill circulation, 259
Surface tension, 3
    of coals liquids, 41
    of hydrocarbons, 41
    of petroleum fractions, 41
    for pure compounds, 41
    and viscosity, 38
Surfactants, 182

T

TAMOC, 13, 14
TAMOC near-field model, 238
Tanker catching fire, 5
Tanker collision, 12
Tanker oil spills
    Exxon Valdez, 93–95
    Haven oil spill, 92
    marine transportation
        allision, 64
        collision, 64
        dead weight (DWT), 60
        Exxon Valdez oil spill, 64
        the Sanchi, 65
        in Strait of Hormuz, 62, 63
        tanker sizes and uses, 61
        very large crude carriers (VLCC), 60
        world oil production, 60
    Persian Gulf oil spill, 95–97
    Sanchi oil spill, 90–91
    of Valdez oil slick, 94
Tanker runs aground, 5
Tankers collide, 5
Tankers towed away, 5
Temperature, 201, 204–208, 215
Temperature-dependent mass transfer
    coefficients, 204
The Atlantic, 7
3D oil spill modeling, 12
Time step, 222
Torpedo, 1
Toxic materials, 149
Toxic oil, 201
Toxicological, 203
Toxicological point of view, 12
Trajectory of an oil spill, 13
TRANSAS, 12

Trinidad & Tobago, 88
Tripoli, Libya, 5

U

UK, 7
UK oil tanker, 2
UK Prime Minister, 7
Unmanned aerial systems (UAS), 269
US Coast Guard, 178
US Coast Guard vessel, 149
US Lawmakers, 7
US oil spill, 106
US president, 6, 135
Uzbekistan, 5, 83

V

Vapor Pressure, 3, 205
Vapor pressure of oil at sea surface
    temperature, 206
Vapor pressure of pseudocomponent, 206
Vaporization mass transfer coefficient, 205, 216, 218
Variation of temperature *versus* time in days, 218
Vertical dispersion, 12
Very large crude carriers (VLCC), 60
Viscosity, 3
    estimation of, 39
    Glaso's correlation, 39
    and surface tension, 38
Volume fraction of components in oil spill, 204
Volume fraction of oil disappeared, 204, 206

W

War, 4
Water salinity, 211
Wax, 202
Weather conditions, 200
Weathering processes
    biodegradation, 255
    dissolution, 254–255
    emulsification, 253–254
    evaporation, 252–253
    photo-oxidation, 256
    sedimentation, 255–256
    spreading, 251
Well blowout, 4
Wellhead blowout, 5
Wetlands, 6
Wide hydrocarbon mixture, 208
Wind speed, 10, 211
Wind speed and temperature in Venice, LA, 218
World energy consumption, 49
World oil production, 50, 60

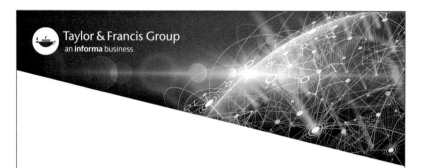